John Ellor Taylor

The Playtime Naturalist

John Ellor Taylor

The Playtime Naturalist

ISBN/EAN: 9783337025908

Printed in Europe, USA, Canada, Australia, Japan

Cover: Foto ©berggeist007 / pixelio.de

More available books at **www.hansebooks.com**

THE
PLAYTIME NATURALIST

BY DR. J. E. TAYLOR, F.L.S.

AUTHOR OF
"OUR COMMON BRITISH FOSSILS," "THE SAGACITY AND MORALITY OF PLANTS," ETC.
AND EDITOR OF "SCIENCE-GOSSIP"

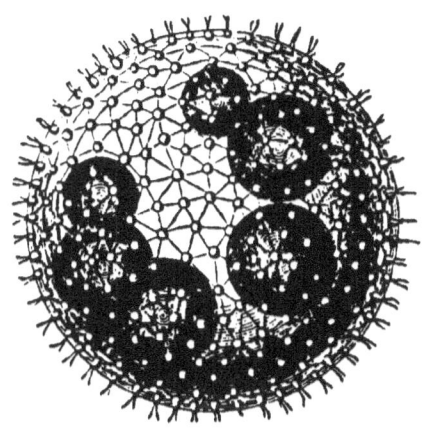

WITH 366 ILLUSTRATIONS

London
CHATTO & WINDUS, PICCADILLY
1889

LONDON :
PRINTED BY WILLIAM CLOWES AND SONS, LIMITED,
STAMFORD STREET AND CHARING CROSS.

PREFACE.

THE writer of this book has a liking for intelligent English lads, just as some people have for blue china and etchings. He ventures to think the former are even more interesting objects. And, as the writer was once a boy himself, and vividly remembers the never-to-be-forgotten rambles and observations of the objects in the country; and, moreover, as he treasures up such reminiscences as the most pleasant and innocent of an active man's life, he thought he could not do better than enlist this younger generation in the same loves and the same pleasures. He has endeavoured to do his best for his human hobbies, and hopes their lives may be richer and sweeter and more manly, for what he has introduced them to in the following pages.

IPSWICH,
December 17, 1888.

CONTENTS.

CHAPTER		PAGE
	LIST OF ILLUSTRATIONS	xi
I.	OUR NATURAL HISTORY SOCIETY	1
II.	FIRST AWAKENINGS	9
III.	AMONG THE BIRDS	20
IV.	NIMRODS AMONG THE LEPIDOPTERA	50
V.	HOLIDAY RAMBLES AND ADVENTURES	83
VI.	LAND SHELLS	124
VII.	"THEY GO A-FISHING"	141
VIII.	A NEW HUNTING-GROUND: AMONG THE MITES	163
IX.	TOADS, FROGS, NEWTS, AND REPTILES	179
X.	SMALL FRY	201
XI.	INVISIBLE LIFE	234
XII.	MICROSCOPIC PLANTS	251
	INDEX	283

LIST OF ILLUSTRATIONS.

A

Achorutes purpurescens (magnified), 118
Actinophrys aculeata, 243
—— eichornii, 249
—— sol, 242
Amœba villosa, with compound pseudopodia, 240
Anatomy of a caterpillar, 104
Antennæ of fresh-water shrimp, 151
Anthomyra pluvialis, 96
Anurea leptodra (magnified), 229
Aphis, winged, 107
——, wingless, 107
Arrenurus, female, 169
—— atax, 178
—— buccinator, 171, 177
—— —— (under side), 171
—— ellipticus, male (upper side), 172
—— frondator, female, 174
—— globator, female, 176
—— ——, male, 176
—— integrator, 175
—— perforatus, male, 170
—— ——, male (under side), 171
—— rutilator, 174.
—— ——, female, 174
—— tricuspidator, male, 173
—— truncatellus, 175

Asellus aquaticus, 152
Asilus crabroniformis, female, 93

B

Bedstraw hawk moth, 70
Blackcap warbler, 43
Black-headed bunting, 36
Black-vein moth, 79
Blind worm, 195
Blowpipe for eggs, 47
Bombylius medius, 95
Bordered white moth, 79
Bramble-leaf brand, 253
Brindle white-shot moth, 77
Bucentes geniculatus, 92
Button galls on oak-leaf, 88

C

Cabinet drawer for eggs, 48
Candle-snuff fungus, 257
Case of caddis-worm, 151
—— Limnephilus flavicornis, 151
Caterpillar of emperor moth, 68
——, cocoon, and imago of small eggar moth, 69
—— of puss moth, 71

xii   LIST OF ILLUSTRATIONS.

Chalk carpet moth, 79
Chrysalis of house-fly, 99
Closterium striolatum, 263
—— Leibleinii, 263
Cocoon of hydrophilus, 147
—— water-spider, 114
Collecting-bottle, 258
—— for diatoms, 260
Colurus deflexus, 231
—— uncinatus, 230
Common house-fly (enlarged), 97
Conops ralipes, male, 91
Corethra plumiformis, 154
Cosmarium margaritiferum, 266
——, empty fronds, 266
Cream-spotted tiger moth, 78
Cristatella mucedo, 219
—— enlarged, showing polypes, 220
Cuckoo, 21

D

Daphnia pulex, 204
—— ——, male, 205
—— ——, female, 205
—— reticulata, male, 206
—— ——, female, 206
Degeeria cincta (magnified), 118
Diagram of larva of gnat, 157
Diphthera orion, 74
Dipper, the, 41
Drills for eggs, 47
Duck-weed, 210

E

Early thorn moth, 76
Egg bag of common gnat, 155
—— drills, 47
—— of buff tip, 58

Egg of cabbage moth, 59
—— of common magpie moth, 59
—— of house-fly, 98
—— of meadow brown butterfly, 58
—— of Pieris brassicæ, 58
—— of Polyommatus corydon, 59
—— of small copper, 39
—— of stone-mite, 165
—— of Vanessa atalanta, 58
Eggs of gnat in various stages, 156
—— newt wrapped in leaves, showing development, 185
—— ranatra deposited on leaves of frog-bit, 150
Emperor moth, 67
End of frond of Closterium lunula (magnified), 264
End of hair-worm, 153
Euastrum didalta, 267
—— margaritiferum, 267
—— oblongatum, 267
—— —— (front view), 266
—— —— (side view), 266
Euchlanis (retracted), 229
—— (exserted), 231
Eyes of spider, 110
—— water-flea, 208

F

First stage in development of hydra, 212
Floxularia cornuta, 228
Foot of Asilus crabroniformis (magnified), 94
Four-spotted footman moth (male), 77
——, female, 76
Fresh-water polyzoon, 217
—— shrimp, 151
Frog-spawn *in situ*, 184

LIST OF ILLUSTRATIONS. xiii

Frog, showing stages of development, 186, 187
Frog-stages of tadpoles, 189

G

Gall insects, 86, 87
Galls on oak-leaf, 88
Garden spider, 111
General form of main track of Hylonomus, 109
Glass tube for sucking eggs, 47
Glow-worm, male, 106
——, female, 106
Goat moth, 64
—— sucker, 44
Golden-eyed gadfly, 91
Gold-shot moth, 79
Great green grasshopper, 117
Group of British lizards, 195
—— Plumatella (enlarged), 223

H

Hair-tailed millipede (magnified), 115
Hairs of Dermestes, 116
—— or feathers of Polyxenes, 115
—— of tail of Polyxenes, 115
Hair-worm, 153
Head of common snake, 196
—— moth, showing eyes, antennæ, and proboscis (magnified), 106
—— viper, 196
Helix aculeata, 132
—— arbustorum, 130
—— aspersa, 129
—— cantiana, 131
—— caperata, 131
—— carthusiana, 132

Helix ericetorum, 129
—— hispida, 131
—— hortensis, 130
—— lamellata, 132
—— lapicidia, 131
—— nemoralis, 129
—— pigmæa, 132
—— pomatia, 128
—— pulchella, 131
—— rotundata, 131
—— rufescens, 132
—— virgata, 130
Herald moth, 77
Hipparchia janira, 57
Hyalotheca dissiliens, 268
Hydra viridis, 211
—— (magnified), 213
—— attacking water-flea, 214
Hydrophilus piceus depositing its eggs, 148
Hyria auroraria, 74

I

Imago of Hylonomous fraxini, 109
—— (magnified), 109
Improvised live-box, 188
—— zoophite-trough, 202
Infusorial parasite of hydra, 239

J

Jaws of Helix nemoralis (magnified), 135-137
Jay, the, 23
Jelly animalcules, 241

K

Kerona polyporium, 243
Kingfisher, 25

xiv LIST OF ILLUSTRATIONS.

L

Lady-bird beetle, larva, and pupa, 106
Lancet of wasp-sting, 103
Lantern and net, 74
Lapwing, 31
Large emerald moth, 77
Larva of beetle covered with compound hairs, 116
—— caddis-worm fly, 150
—— Dytiscus marginalis, 145
—— goat moth, 63
—— Hylonomous fraxini, 109
—— Micropteryx subpurpurella, 62
—— Nepticula durella, 62
Leaf-cutter bee cutting piece out of leaf, 90
Lepisma (magnified), 122
Leptogaster cylindricus, 95
Liparogyra dentreteres, 273
Lithosia quadra, 75
Long-tailed tit, 27

M

Maggot of house-fly, 98
Maple blight, 254
Mastigocerca bicristata (magnified), 232
Meadow-sweet brand, 252
Melicerta ringens, 224
Micrasterias rotata, 265
Mined bramble-leaf, 61
—— oak-leaf, 61
Mite from Gamasus of humble-bee, 167
Myopa testacea, 92

N

Narrow-bordered clear-wing, 53

Natterjack toad, 194
Navicula didyma, 279
Nest of dipper, 41
—— reed-bunting, 37
—— spider, 112
Nitzschia vivax, 274
Nuthatch, 42
Nymph of gnat, 157

O

Oak hook-tip moth, male and female, 76
Orthosira Dressæri, 274
Ovarium of fresh-water sponge, 248

P

Pale oak beauty moth, 76
Paludicella sultana, 221
—— (enlarged), 222
——, showing polypes, 222
Parasite of Dytiscus, 146
Philophora plumigera, 78
Phyllactidium pulchellum, 271
Pennularia borealis, 275
—— major, 278
Pleurasigma formosum, 279
Podura without scales, 119
Pupa of goat moth, 64

R

Ranatra linearis, 149
—— catching its prey, 149
Red-belted clear-wing, 55
Rose-leaf cut by leaf-cutter bee, 90
Rotate or wheel-shaped spiculæ, 249
Rotifer vulgaris (magnified), 229

LIST OF ILLUSTRATIONS.

S

Sand-lizard, 199
Scale of black Podura, 120
—— bleak, 12
—— bream, 14
—— carp, 16
—— chub, 11
—— dace, 13
—— eel, 12
—— grayling, 18
—— gudgeon, 14
—— Hipparchia janira, 57
—— loach, 15
—— minnow, 15
—— perch, 16
—— Pieris brassicæ, 56, 57
—— pike, 17
—— Polyommatus alexis, 56
—— roach, 13
—— speckled Podura, 120
—— Vanessa urticæ, 55
Scarlet tiger moth, 75
Section of button gall (magnified), 89
—— diatom commencing deduplication, 276
—— spangle (magnified), 89
—— sycamore-leaf, 255
—— viper's head, 197
Sedge-warbler, 35
Selidosema plumaria, 75, 79
Setting-board for Lepidoptera, 80
—— out Lepidoptera, 80–82
Side view of zoophyte-trough, 202
Single eggs and young of Ranatra, 150
Small black arches moth, 74
—— emerald moth, 79
Smooth newt, female, 191
——, male, 192
Smynthurus niger (magnified), 122

Spangles on oak-leaf, 88
Speckled Podura, 121
Spinneret of garden spider, 111
—— gossamer spider, 112
Spirogyra in different stages, 270
Spores and cells of "witches' butter," 256
Stages in development of Epistylis, 247
—— Euglena viridis, 244
—— fresh-water snail, 139
—— Stephenoceros, 225–227
—— of metamorphosis of Pieris brassicæ, 65, 66
Star-spored brand, 252
Statoblasts of Plumatella developing, 219
Staurastrum dejectum, 267
—— alternaus, 267
—— gracile, 268
—— spongium, 268
Stauroneis phœnicenteron, 278
Stictodiscus Californicus, 277
—— gracile, 268
Sting, lancet, and poison-bag of wasp, 102
——, poison-bag, and poison-gland of humble-bee, 100
Stings of hydra, 215
Sycamore - leaf with Melasmia agerina, 255
Synchæta longipes (magnified), 230

T

Tadpole of frog, 188
Teeth of blow-fly (magnified), 105
Tegenaria atrica, 110
Terminal spiracle of Dytiscus marginalis, 146
Tetranychus lapidus, 165
—— populi, 166

Tetranychus salicis, 165
—— telarius, 164
—— tiliaris, 164
—— ulmi, 165
—— urticæ, 167
—— viburni, 166
Theridion riparium, male and female, 110
"Thousand legs," 108
Thyatira batis, 76
Toad, the, 190
Tongue and lancet of common flea, 103
Track of Hylonomus fraxini beneath the bark of a tree, 109
Transparent burnet moth, 55
Tunic of dead polyp filled with stato-blasts, 218

U

Ulothrix, 269
Umbrella net, 63

V

Vaginicola before and after fission, 238
Vapourer moth, female, 72
——, male, 72
Volvox globator, 271
—— stellatum, 272
Vorticella nebulifera, 246

W

Water beetle, male and female, 144
—— flea, female, 203
—— spider, male, 113
—— ——, female, 114
Wheatears, 33
Winged aphis, 107
Wingless aphis, 107
"Witches' butter," 256

Y

Young hydra, 215
—— of Synchæta longipes, 230

THE
PLAYTIME NATURALIST.

CHAPTER I.

OUR NATURAL HISTORY SOCIETY.

JACK HAMPSON was a capital sample of the best traditions of Mugby School. A lad of fourteen, with well-knit limbs, brave, honest-looking, bluish-grey eyes, a good cricketer and swimmer, and not bad at a high jump. He could no more do a mean thing than he could tell a lie; and he could give or take a thrashing if absolutely necessary, although he would be in no hurry for either.

Mugby School has kept the lead in modern educational progress which a former distinguished master introduced many years ago. That master was not content that boys should learn Latin and Greek. He was more anxious they should learn to be Christian gentlemen; to fear and eschew an untruth as they would poison; to be brave and yet gentle; tender towards the weak, not defiant even to the strong. The boys at Mugby School were well acquainted with the lives of the best men of all ages and of all nations, as well as with the most stirring deeds of valour, self-denial, and manly bravery. The noblest thoughts of the wisest men were drawn freely upon for their benefit.

Much of this "new education" was thought an innovation at first; but never before were English lads turned out of school in such high-toned, manly form, or so well able to hold their own at the universities, or in the bigger world outside.

As may be imagined, the wonders of science had not been ignored in such a school. One can hardly believe that modern science is almost included within the present century. All before then, except astronomy, was more or less speculation. Nobody would call Linnæus's system of botany a science, although it was very useful and introductory; nor was geology, zoology, nor chemistry. Scientists had only been playing, like children, in the vestibule of the great temple. It may be that we ourselves have not advanced far within the precincts—at least, those who study these subjects a hundred years hence may think so. But, at any rate, the amount of knowledge extant concerning the world in which we live, and its ancient and modern inhabitants, is vast compared with what it was when the present century commenced.

At Mugby School, science was an important and also a welcome subject. How welcome it was is best indicated by the fact that the boys got up a Natural History Society among themselves. This was really a self-imposed task, done out of school-hours. Some of the principal teachers encouraged the lads by becoming members; not that they knew much of natural history or scientific subjects (some of them, indeed, knew nothing at all, and actually learned a good deal from the boys themselves).

Of course, the Society was founded on the best models. It was not a bit behind the famous "Royal Society of London" in its equipment. It had its president and vice-president, and its committee were called "the council." It also published, for the world's benefit, abstracts of the short papers the boys read—the abstracts being nearly as long as the papers. Although its members were not numerous, they felt they bore the weight of the dignity of the Society on their shoulders; and, as they were too boyish-manly to be priggish, the training did them no harm.

Well, the Society was divided into sections.

One section was appointed to collect the plants of the neighbourhood—that is, those obtainable during the school half-holidays; another to collect butterflies and moths; a third, beetles; a fourth, birds; a fifth, fossils, etc. They were to publish lists of the plants, birds, insects, and fossils of the district in the " Society's Proceedings ; " for, of course, the latter was the name given to the abstracted papers.

The Society had only been founded the year before Jack Hampson was sent to Mugby School; so it was in the first zeal and freshness of its youth. Jack didn't like science—it was nothing but a lot of hard, jaw-breaking names, he said, and what was the good of them ? He and others had enough of hard words in their daily Latin and Greek tasks. Jack rather snubbed the fellows who volunteered to learn more hard words than were required—he couldn't understand it. What was the good of calling a buttercup *Ranunculus*, and a white stone *quartz?* It was all sham and show!

Now, Jack was a born hunter. He was ardently fond of fishing, and not a bad shot, considering he had been mistrusted, instead of trusted, with a gun. I dare say his skill with the latter would have astonished his father; and I have no doubt a good many ounces of 'bacca found their way into the keeper's pocket before he became so creditable a shot.

But there was not much fishing about Mugby ; or, rather, they were such little things that Jack felt ashamed of pulling them out, and so he slipped them in again, although they never seemed to grow any bigger. This was a wise act on their part, if they had only known the unconscious chivalry of Jack's nature, which hated taking advantage of a weak thing. Then as to shooting—first, he hadn't a gun, and if he had possessed one, the rules of the school would have precluded his using it. Next, what was there to shoot? The small birds in the hedges? Any cad could do that! Sneak after the poor beggars behind hedges, and then bang at a robin, a wren, a yellow-hammer, .or a tit, and perhaps blow it to pieces! That was not good enough. Partridge and pheasant shooting, Jack thought, are hardly much better sport, only you can eat them.

Of course, there was the excitement of cricket and football, hare-and-hounds, paper-chases, hurdle-racing, jumping—not only not bad, but altogether good and brave and manly sports. But, somehow, a lad of superior mental abilities wants something else.

Now, the scientist is also a hunter. He traces his descent from Nimrod—he is a hunter before the Lord. He roams through the stellar universe for his prey—hunts for stars, comets, planets. He is not daunted because he did not live on the world when it was young, millions of years ago; for he makes up for it by hunting the remains of the animals and plants that lived during countless ages, and which have long been buried in the rocks of the earth's crust as *fossils*. He hunts for flowering plants and animals in all parts of the earth; braves heat and cold, hunger and thirst, wounds and death, in his ardent search for them. The structures of rocks do not escape his mineralogical hunting, nor the composition of any sort of substance, organic or inorganic, his chemical analysis. He hunts down stars thousands of millions of miles away with his telescope, and creatures less than the fifteenth-thousand part of an inch long with his microscope. Was there ever such a great hunter? This hunting instinct began scores of thousands of years ago, when the hairy, naked Palæolithic men hunted extinct hairy elephants and rhinoceroses. It has been developed until it has assumed the high intellectual pleasure of roaming through God's great creation, and of confirming the ancient writer's conclusion—"Lo, there is no end to it!"

Of all these things Jack Hampson had never heard a word. Perhaps he had occasionally listened to a few joking remarks about Darwin and our "being descended from monkeys" at his father's dinner-table. But his father (who was anything but a wealthy man these hard agricultural times, although he farmed his own estate) had not much time for considering the discoveries of modern science. Their echoes faintly reached him occasionally, but never touched him seriously. Not only were the times bad, but his family was large, and it was not without a stretch that Jack was sent to Mugby School, rather more than twenty miles off. His brother (Jack's uncle) was better off, because he had no family; and the uncle also had more leisure, and, what is more, was really a man of a literary and scientific turn of mind.

All schoolboys make friends at school. Nobody has ever analyzed the process of friend-making among boys. It is as mysterious as genuine love-making. Friendships—at least, boys' friendships—are also made "at first sight." Live in a public school a few years, and you will find it out. You might just as well tell a boy to make friends with a certain other boy, as order him to make love a few years later with your female selection! And yet what issues of life depend on those boyish friendships made at school! They are often more durable than marriages. They survive success, disaster, and disease. Not unfrequently, they are prolonged to the second and third generation. If there is one thing more difficult to explain concerning instincts than another, it is the instinct of boys' friendships.

How Jack Hampson—big-limbed, broad-backed Jack—came to take up, the very day he arrived at Mugby, with little Willie Ransome, I cannot tell. There is something in the doctrine of contrasts; doubtless Willie was as great a contrast to Jack as you would have found in the whole school —rather undersized, weakly, but nevertheless a brave and truthful boy. He was fond of books— a trifle too fond, for it would have done him good to have got away from them a little. The chief feature about Willie was his large, bright, inquiring eyes, and his altogether affectionate disposition. He took to Jack at once, and Jack to him. Never before was there a better illustration of "friendship at first sight."

CHAPTER II.

FIRST AWAKENINGS.

IT was at the commencement of the Spring Term that the friends came to Mugby School. Without knowing it, but fortunately for them and for the whole school, a fine enthusiastic young fellow had been appointed "science teacher." The term sounds vague, but so do all terms if too strictly analyzed. The boys dubbed him "professor," and thereby unconsciously gave him higher rank than his *confrères*, who were only "teachers." It would have been impossible for a young man to have been selected better fitted for such a post. Nothing gets hold of boys sooner than enthusiasm. Boys are naturally enthusiastic. There is no better proof of vitality even in an old man, than that he continues to be enthusiastic about anything intellectual.

Willie Ransome's father was a village doctor, and it was hoped Willie would some day help his father in his increasingly larger, but not increasingly profitable, rounds. Willie entered the science class the first term. His father was a man of scientific tastes, with little leisure to indulge them. But he had already inoculated his only son with a love for such subjects. Willie, however, had never before been drawn within the magic circle of enthusiasm for them, and his highly sensitive temperament was fixed by the professor's descriptions and demonstrations immediately. Before the term was half over, he was a member of the Society, and doing his best to "collect" for the Society's museum.

Jack had many a hearty laugh over this disposition to hoard up a lot of old stones and things, and give them hard names. More than once he was asked to attend a Society's meeting—for each member had the privilege of introducing a friend—but he always shirked it. "No," he said; "they are not my sort."

One wet evening, however, Willie Ransome got Jack to go, just because there was nothing else to do. There was a short paper being read on "Fish Scales," and a number of them were mounted for microscopical examination, of course with a low power, say inch and half-inch. Anything relating to fish or fishing was certain to gain Jack's attention, therefore a better subject could not have been selected to engage his notice. Besides, Jack had never yet even looked through a microscope! He felt a bit ashamed of this now; but there were a couple of microscopes present, and Jack determined to have a good look through them. The scales of

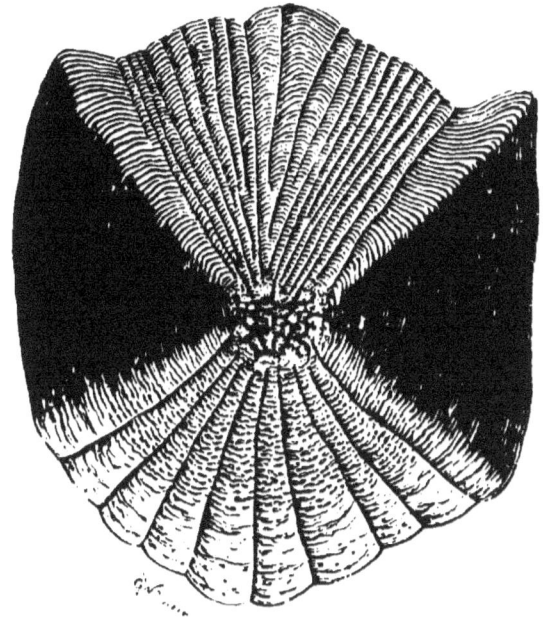

Fig. 1.—Scale of chub.

different sorts of British fishes were on view. Of course, fish-scales are common enough; but who would think that each kind has its own pattern of scale, and that you could tell a species of fish by its scales?

The paper showed that the scales of fishes were

composed of the same material, *chitine*, as the feathers of birds, or the hair and nails of animals—

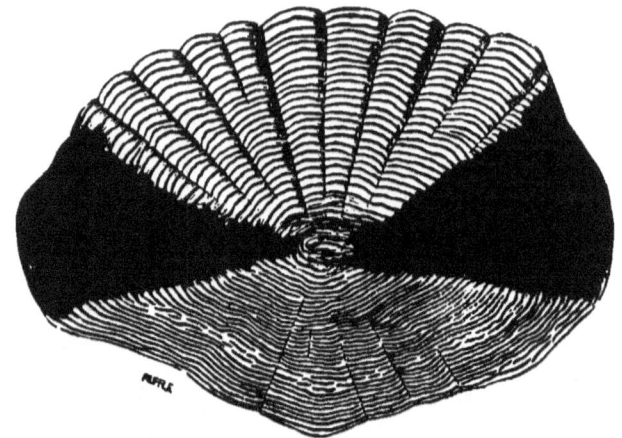

Fig. 2.—Scale of bleak.

a kind of substance only found in the animal kingdom, and never in the vegetable; that these scales

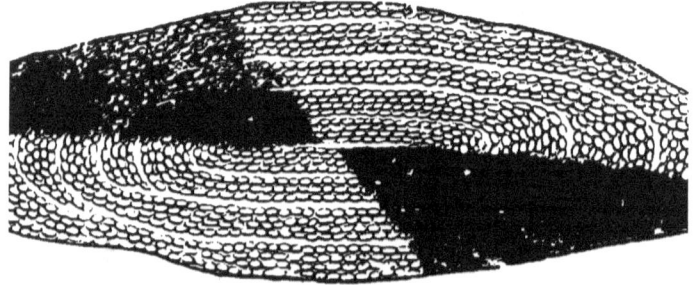

Fig. 3.—Scale of eel.

are developed in little pockets in the fish's skin, which you can plainly see for yourself when a herring is scaled. They are arranged all over

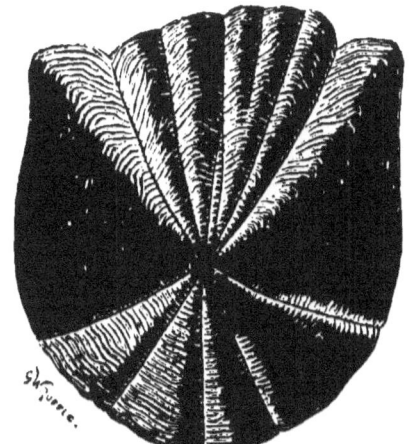

Fig. 4.—Scale of roach.

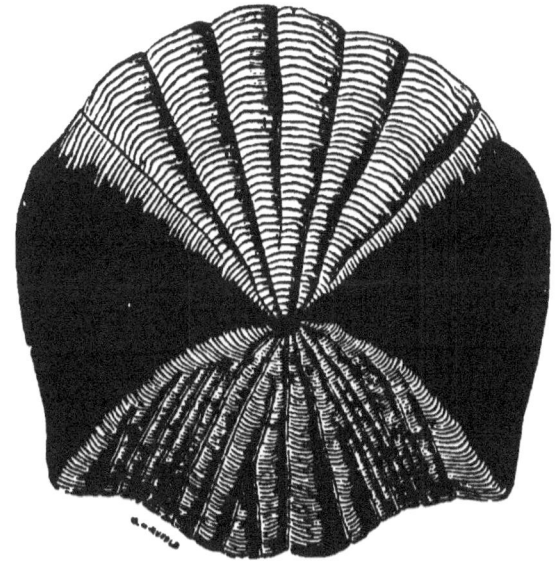

Fig. 5.—Scale of dace.

the fish's body like the tiles covering a roof, partly

14   THE PLAYTIME NATURALIST.

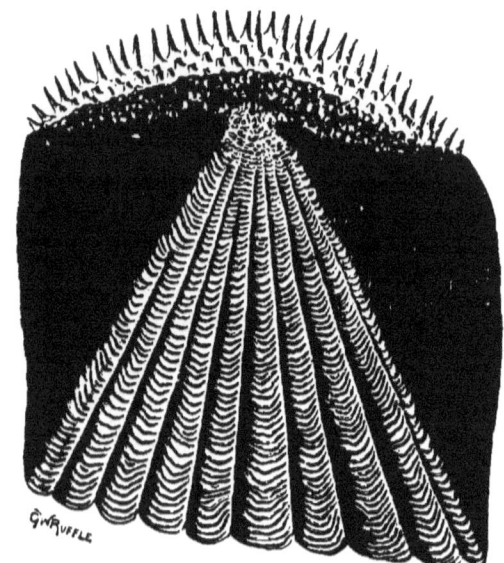

Fig. 6.—Scale of gudgeon.

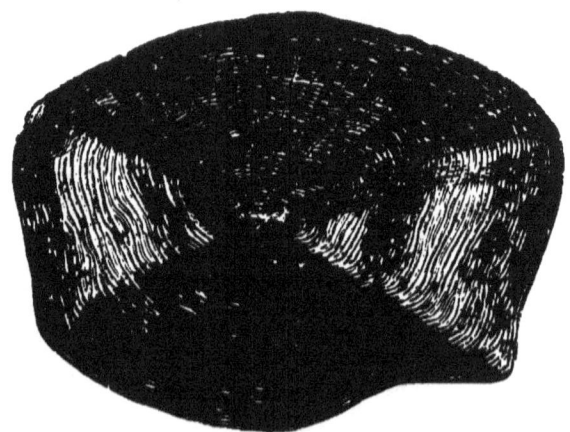

Fig. 7.—Scale of bream.

overlapping each other, as is seen by one part of the scale being often different from the other.

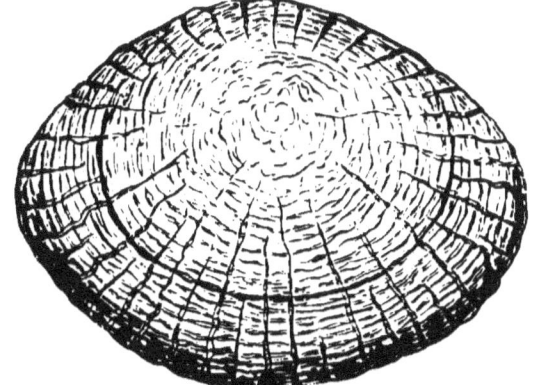

Fig. 8.—Scale of loach.

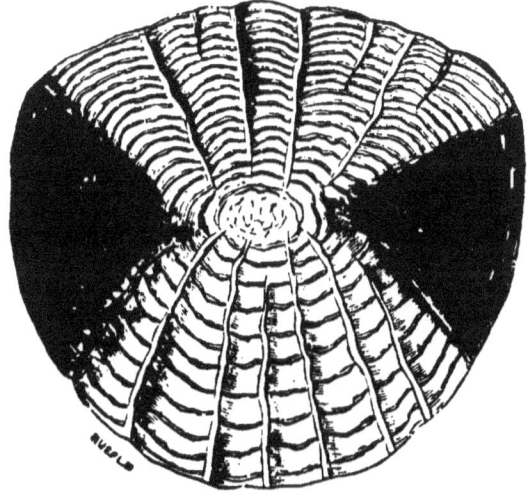

Fig. 9 —Scale of minnow.

Jack looked through the microscope, and was delighted. He was always a reverent-minded boy,

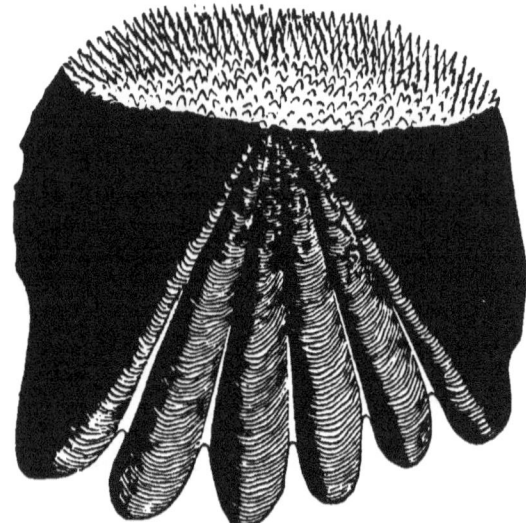

Fig. 10.—Scale of perch.

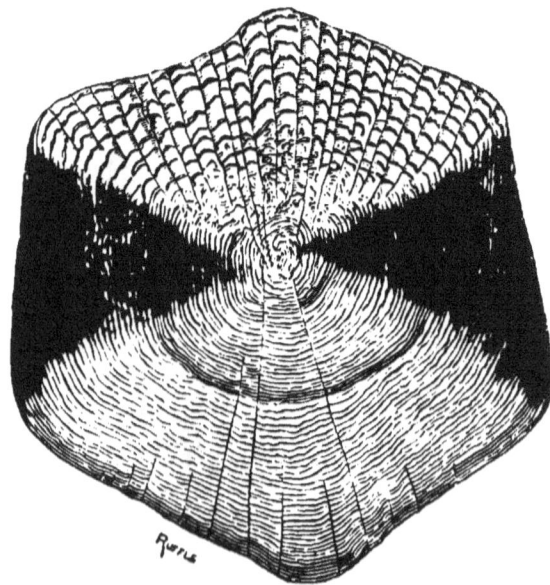

Fig. 11.—Scale of common carp.

FIRST AWAKENINGS.

and the sight broke on his mind like a new revelation. How exquisitely chaste and beautiful were the markings, lines, dots, and other peculiarities! Then the scales which run along the middle line

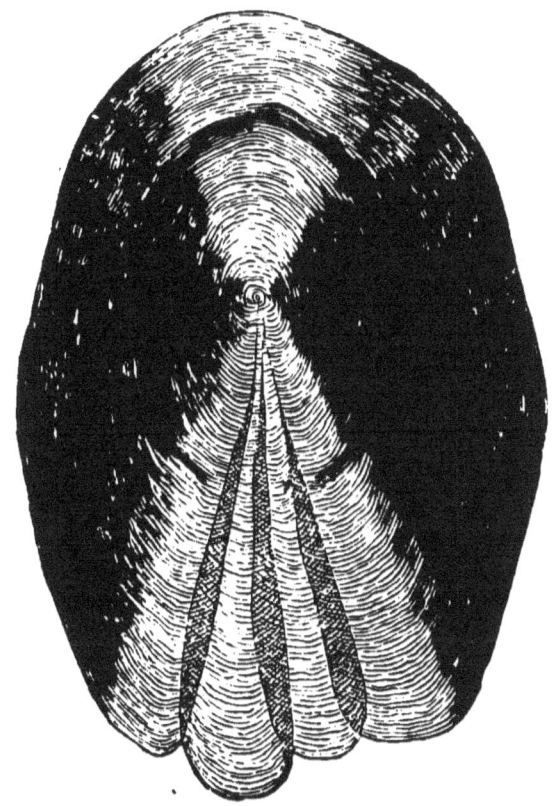

Fig. 12.—Scale of pike.

of the fish were shown him, and the ducts perforating them, out of which the mucus flows to anoint the fish's body, and thus reduce the friction of its rapid movement through the water. The lad was half bewildered at the possibility of the new knowledge. "Could anybody get to know about these things?" he asked Willie, who told him of course he could, if he ould only take a little trouble.

"But," said his young friend, "I would advise

Fig. 13.—Scale of grayling.

you to get a pocket-magnifier first, and begin to examine with that. Some fellows begin right off with a powerful microscope they get their governors to buy them, and they work it like mad for a month or two, and then get tired of it. Fact is, they never learned the art of observing."

"What do you mean by that?" said Jack.

"Why, getting into the habit of looking about you, keeping your eyes open, and quickly spotting anything unusual. Fancy a fellow beginning to use magnifying glasses of thousands of times before he has begun to use his own eyes! Use your own eyes first, then get a little extra help in the shape of a shilling pocket-lens, and by-and-by you will be able to use a real microscope, and enjoy using it too."

This was rather a long lecture for Willie to give, or for Jack to listen to. He wouldn't have listened if it had not been for what he had just seen. He said nothing, but he made up his mind he would get one of these useful shilling magnifiers. Willie usually had a country walk during the school half-holiday, and Jack had often been invited to accompany him; but he didn't care to go "humbugging after grubs and weeds," he said. Now, however, he invited himself, and somewhat surprised his friend by stating he wanted to go with him.

CHAPTER III.

AMONG THE BIRDS.

IT was a bright afternoon in early summer. The hedges and woods were full of bird-music. You couldn't see many birds, for the luxuriant foliage screened them, but there they were; a hundred pairs of bright birds' eyes watched the young friends as they sauntered along the shadow-flecked roads. Overhead the lark was raining down its melody. That "wandering voice," the cuckoo—the Bohemian among British birds—was heard, in the first freshness of its call-note; for, as the proverb goes in Suffolk—

"In May he sing all day;
In June he change his tune."

Whether it be true or not that the female cuckoo has the power of changing the colour-tone of her eggs, and adapting them, as a sort of mimicry, to the colour of the eggs in the nest into which she surreptitiously slips her own, has been a disputed point. But one thing is certain—the cuckoo has a marvellous power of modifying the colour and even markings of her eggs. You can hardly find two eggs of the cuckoo marked and tinted exactly alike. This restless bird appears to have drifted away from its

Fig. 14.—The cuckoo (*Cuculus canorus*).

oological moorings. The cuckoo is the only British species. North America appears to be its headquarters. There and elsewhere cuckoos build nests like other birds, and have regularly marked eggs. In this country, the female cuckoo is the victim of polyandry—she has too many husbands! They don't give her time to build a nest and attend to her domestic duties, like other birds. So she has become a by-word and a scorn among the chaste avian matrons who may be seen following her any May morning, as if she bore on her breast the "scarlet letter."

In that hazel copse on the right, where the nightingale is trying its early notes, you hear the harsh grating cry of the jay. It is getting quite a local bird now, which is the first step towards its becoming a rare one. Our game laws have had an important influence on our native zoology, and even botany. Every creature which an ignorant gamekeeper regards as injurious to the birds and eggs under his charge, is condemned to death. Consequently there are few mammals or birds which he does not regard suspiciously. The gamekeeper's idea of the proper fauna to inhabit the earth is—first, pheasants, then partridges, next, hares, and (a long way behind) rabbits! Why Providence created anything else is a mystery to him, and tries his bump of reverence sorely.

The sweetly pretty blue which glances in the

wing-feathers of the jay has been against its prosperity. Of course, they were developed to please—*jays*. But in these later times they have pleased human beings of the female gender, and that is a

Fig. 15.—The jay (*Garrulus glandarius*).

bad thing for pretty birds. Women, and especially *young* women, all over the world, labour under the mistaken idea that they are not good-looking enough—that a few pretty feathers torn from the wings of pretty birds, shot and maimed and robbed of their brief lives for the purpose, would render them yet more attractive! The mistake is unfortunate for the birds.

We have not many birds whose *colours* attract attention in our sober British Isles. True, the kingfisher is still common among us, thank Heaven! You may yet see it flash past like a sapphire, even in winter. The idiots who can afford to pay the gun-tax, and who have just got sense enough to kill something or hurt something for life, have not yet been able to shoot down the kingfisher. Male and female are almost alike in their rich cerulean, prismatic plumage, thanks to the fact that the female nests in a *hole*, which thus conceals her lovely colours whilst she is sitting. Mr. A. R. Wallace has shown that in most cases where the female is as brilliantly coloured as the male, the nest is concealed. A brilliantly coloured bird, sitting for two or three weeks, would be a conspicuous mark to her enemies if her nest were an open one. Hence the reason why the female pheasants are so dull-coloured, whilst the males are so brilliant.

A funny nest is that of the kingfisher, when you find it—rather badly built of interlacing fish-bones instead of grass and hay, or moss; but not an inartistic structure nevertheless. It seems a strange way of utilizing your waste food—to construct your lodgings out of it!

Fig. 16.—The kingfisher (*Alcedo ispida*).

The two lads were more silent than lads usually are on an exuberant morning like this. The fact was, both were genuine naturalists without knowing it. The true naturalist is a true poet. Into his mind the influences of natural scenery, of natural history, unconsciously sink down. There is an unmentionable bliss in the unrecognized sympathy which goeth forth towards all things into which He hath breathed the breath of life. The scents of the opening buds—too fragrantly evanescent even for the cleverest *parfumeur* to fix —the hallelujah chorus of summer voices, birds chiefly, but not only, which enter the "Emanuel's gate" of the human ear; the sad, soft sighing of summer winds; the unobtruding kaleidoscope of floral form and colour, scattered so freely and bountifully;—cannot these get hold of the soul of a man? One feels constrained to adopt the language of the principal talker among the favourite disciples—"Lord, let us build three tabernacles," etc. The disciple was in no hurry to depart.

Just after, the boys—who had enjoyed each other's speechless company, until they began that pastime common to boys of all characters all over the world, *nest-finding* — happened to stumble across perhaps the most remarkable nest of all our British birds—that of the "pudding-poke," or long-tailed tit (*Parus longicaudatus*). In the old hawthorn hedge, covered with grey and yellow lichens, the long purse-like nest was so externally adorned with similar lichens that you could with difficulty tell the nest from the lichen-clad fork in which it

Fig. 17.—The long-tailed tit (*Parus longicaudatus*).

was fixed. Never was a cleverer bit of *mimicry*, or pretending. Lads whose play and pastimes incline them to be Indian chiefs, brigands, pirates, robbers, etc., can appreciate this pretending, or *mimicry*, on the part of birds and insects perhaps better than their elders. They do it every day for sport; the poor birds and insects do it every day in earnest, for their lives and specific existence depend upon it.

But there is a comical side even to the most serious engagement in life, if you care to seek for it. Here is the long-tailed tit, for instance—what in the world is the good of that long tail to him? He can't use it; it isn't an ornament; and yet he is as proud of it as if it were a peacock's. He goes to the trouble to laboriously construct a long "pudding-poke" nest, simply because he wants room for that useless long tail! Was there ever anything so absurd? Some persons imagine that in this world it is only given to men and women to make fools of themselves. The long-tailed tit also gets a chance.

But of all the tits, give me the common blue tit. That bird is a source of comfort and delight to me all the winter through. He comes to the bone I hang from the bough of the pear-tree in front, of my dining-room; and it is capital fun to see him climb down the string, with all the sparrows sitting around on the nearest boughs, wishing they could do the same, and glad to pick up the crumbs which fall from this lucky bird's table.

Willie knew all these common birds. Their songs were as familiar to him as his own language, from the melancholy alarm-cry of the nightingale to that of the blackbird.

The metallic notes of the chaffinch are heard from every tree. That bird was now in his gayest and neatest plumage, and the male was not at all unwilling to show off his recently acquired plumage. The male yellow-hammer, also, was nearly the same colour as a canary. Before long you hear this bird all along the roads and lanes, uttering that remarkable plaintive cry which has obtained for it in Suffolk (the intonation of whose dialect it somewhat resembles), " A-little-bit-o'-bread-and-*no*-cheese!" Birds' names are frequently *onomatopœic*, or founded on the cries they utter and the sounds they make. Thus the chaffinch is commonly known by the name of "spink," because of the flat, metallic note somewhat resembling this sound.

There is a bird-language, as true ornithologists are aware—a language expressive of joy, as in the song of the ascending lark, and that of the thrush sitting on topmost boughs in the early summertime; of alarm; even of humbug and deceit, as when the lapwing tries to decoy you away from her nest by sham cries of pain. This noble bird was now in view, for the boys had come to a break in the wooded lanes, where a patch of yellow gorse-clad common relieved the pleasant monotony of greenness.

It was a glorious sight, that wilderness of gorgeous yellow. The tradition ought to be true, even if it is not, that when Linnæus first saw the gorse in full flower, he thanked God for allowing him to see a sight so beautiful. I have never beheld, even in the tropics, anything equal to it, much less to excel it. The only approach to it in floral beauty is the wild Australian bush, where the pink epacris take the place assumed by our English gorse.

The gorse is also known as "whin" and "furze," according to the locality. Those breezy birds, the whin-chats, were abroad, flitting from gorse-bush to gorse-bush, resting for a few moments on the topmost branches ; then, flicking their tails and uttering their short, sharp notes, they went away a few yards farther. Like the beautiful lapwings, they do this to lure one from their nests.

The lapwings were wheeling and crying all over the place. Half a dozen couples made more row now than a flock of hundreds of these birds would have done in winter. Their "pee-weet" cries have obtained for them this additional name.

But all their fuss was to lure the lads away, and the

Fig. 18.—The lapwing (*Vanellus cristatus*).

birds fussed and cried all the more when they found

their artifices succeeding. Then, when the dangerous trespassers had wandered a sufficiently safe distance, away flew back the deceitful birds to their nests and young, to practise the same device again when the next trespasser arrived.

All this and more Willie pointed out to his friend, who, indeed, was not entirely ignorant of birds and their ways. Few boys who live in the country are.

Coming to a kind of gravel and sand pit, they were about to go in, when they saw a couple of wheat-ears frolicking about—birds with a remarkable mixture of daring and timidity in their characters. Of course, the face of the sand-cliff was drilled with scores of holes made by the sand-martin (*Hirundo riparia*), which well deserves its specific zoological name, for these communities are really bird-cities in the sand-bank.

Beyond the common, the land sank into a marsh. It was a capital hunting-ground. That queer insectivorous plant, the sun-dew, grew abundantly —indeed, the ground was of a brick-red tint in places with its pretty rosette-shaped groups of leaves—every leaf an ingenious little fly-trap. The place was pink with the lovely flowers of a plant which deserves a better name than louse-wort

(*Pedicularis*). A stream ran through the middle of the little marsh, and rapidly made its way through

Fig 19. —Wheatears (*Saxicola œnanthe*).

a miniature gully it had cut for itself a quarter of a mile further away. Groups of tall trees lined the path of the stream, as it slowly sped through the marshy places; and here, at night, you could

Fig. 20.—The sedge-warbler (*Acrocephalus phragmitis*).

hear the sedge-warbler (*Acrocephalus phragmitis*) feebly imitating the nightingale. Indeed, across the sea it is known as the Irish nightingale. It is one of the cleverest of our British mocking-birds, and can imitate the songs of the thrush, lark, and even its fellow-companion in the same habitat, the reed-sparrow, or reed-bunting (*Emberiza schœniculus*). The boys soon set to work to hunt for the

Fig. 21.—Nest of sedge-warbler.

nests of these reed-birds. After a good deal of sloppy tumbling about, they found that of the sedge-warbler—an exquisitely constructed and ingeniously concealed bit of bird-architecture. It was not long before they saw that of the reed-bunting, a still

36   THE PLAYTIME NATURALIST.

more remarkable construction. It is now too much sought after by those æsthetic people who have suddenly found out that peacock's feathers and sunflowers are beautiful objects, and who owe a debt of gratitude to Japanese artists for drawing their

Fig. 22.—Black-headed bunting (*Emberiza melanocephala*).

attention to the graceful shapes of reeds and grasses. In the drawing-rooms of such people you will see great pots of bull-rushes, and perhaps a cluster of water-reeds (*Arundo phragmites*), with a reed-bunting's nest in the middle.

Notice how cleverly—as cleverly as the Indian weaver-bird—the reed-bunting has twined the grass

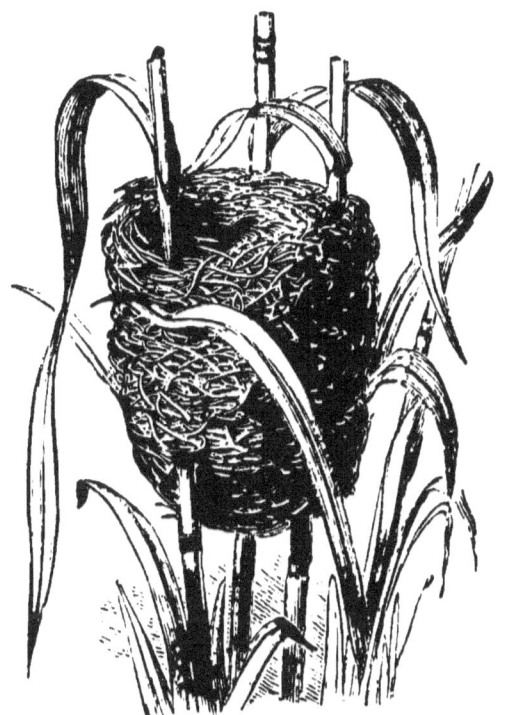

Fig. 23.—Nest of reed-bunting.

structure of its nest in and out of the tripod of three strong reed haulms. The wind may rock it to and fro, but the nest is safe enough. No land animals, weasels, cats, etc., can get to it. Then, as the boys found out, the nests were always made in the reeds lining the edge of the running water. It seemed impossible to get to them except by boat, and there was no boat hereabouts; even if there had been, I doubt whether it would have been used on that small swampy stream.

But the reed-bunting's nest had to be got; not for any æsthetic purpose, but solely because it seemed impossible to get it. That surely is reason enough for a British youth! The bunting's nest could be approached within half a dozen yards, but that half-dozen was a Serbonian Bog. The tall reeds whistled like pan-pipes (the Greek god made his musical instrument out of them); the sword-like blades of the yellow iris ("Fleur-de-luce" or *Louis*—the charge, or golden lilies, on the white flag of royalist France) were topped by the flower-spikes. A few flowering-rushes (*Butomus umbellatus*) were sprinkled among them. Margining the stream were arrow-heads (Sagittaria), water-plantains (Alisma), and other aquatic plants.

But these things were regarded as naught. It was the reed-bunting's nest that was required. Of course, it could have been got at with a long plank, but boys don't carry long planks about with them when they go bird's-nesting. There was a short and decisive council of war held; the result was a stripping of habiliments. Jack was soon in his own skin, plunging cautiously through the swamp; then, as the boggy mass became more watery, lying down on it, wriggling through it like an eel, until at length the reed-bunting's nest was reached. The proud victor over a difficulty returned with his prize, if not a sadder, a differently coloured boy, for the black mud had made him as piebald as a magpie.

That, however, was only part of the fun. A few hundred yards lower down, the swamp ended, and the stream flowed rapidly towards its little gorge. The piebald young Briton hastened thither, and plunged into as deep a hole as he could find, to wash off the stains of his recent campaign. In doing so he startled a couple of those remarkable birds, the water-ousels, or dippers (*Cinclus aquaticus*), a bird which has been much hunted down in salmon and trout streams, on the alleged reason that it destroyed their ova. Nothing of the kind. It simply haunts streams for the sake of caddis-worms and other insects, shell-fish, etc. Perhaps it does vary this monotonous diet by a little fish, very little fish—small fry, in short.

There was a suspiciousness about the water-ousels, when they were startled, which led Willie to

Fig. 24.—The dipper (*Cinclus aquaticus*).

believe they had a nest thereabout. So, as soon

as Jack had got rid of the Ethiopian part of his skin, they set about to look for it. At last it was found, and a pretty, comfortable little dwelling it was. The lads left it, for it was a pity to take such a nice little bird's house. Besides, there was no adventure connected with taking it, which goes for a good deal in a boy's moral code.

Fig. 25.—Nest of water-ousel.

The afternoon had well set in now, and the young friends were sharp-set as to their appetites. It would be a grand thing if there could be invented a sort of *proventriculum*, or paunch, for young lads who wander miles to study natural history and procure specimens ; then, if they felt hungry, they could refresh themselves.

Nevertheless, as they came to the wooded lanes and roads again, they were not too hungry to fail

Fig. 26.—The nuthatch (*Sitta Europæa*).

in catching the nuthatch at work—small, insignificant-looking bird though it is. The blackcap (*Curruca atricapilla*) was warbling as I verily believe only blackcaps can. "Warbling" is the best word wherewith to designate its song. A cockney authority on music and morals has put it on record that there is no real music in the songs of birds! It is a pity they cannot sue him for libel. Nobody but a cockney would have uttered it. Shakespeare did not think so, nor did Shelley or Wordsworth,

Fig. 27.—Blackcap warbler (*Sylvia atricapilla*).

and the world values their opinions almost as much as those of Mr. Haweis.

On their way back to school—tired, hungry, silent—they heard, through the deepening gloaming, the "churring" cry of the goat-sucker (*Caprimulgus Europæus*), sometimes as close to them as the bats;

at others, far away in pursuit of the larger moths or beetles. That *jarring* note has procured for this

Fig. 28.—The goat-sucker (*Caprimulgus Europæa*).

bird its other name of "night-jar;" whilst its love for ferny, bracken-clad slopes (especially where pine woods break their monotony with covers) has given it the other British name it bears, especially in the midland counties, of "fern-owl."

Is there anything more delightful than the fatigue of a summer afternoon's long ramble after objects one loves? You are not tired of them, but *with* them. It is a delicious fatigue. Subsequent years of trouble cannot obliterate the charmed impressions. They are the sunniest spots in one's memory. Their recollections come, like angels' visits, to unconsciously relieve us in after-years of many a sad trouble and trial. They should be laid up in store when you are young, so that they can be drawn upon when you are old. Then the sunshine of youth is stored to gild the troubled days of matured manhood and the darker shadows of old age.

Next day, the few eggs the boys had collected (and in collecting them they had taken conscientious care not to interfere with the clutches unless they were full or nearly so, in order not to disturb or interfere with the bird's laying) were all laid out on Willie's little table. He had the handy little volume entitled "Collecting and Preserving Natural History Specimens," and had turned to Mr. Southwell's capital chapter on collecting and arranging bird's eggs. The young friends read it through together. The delight of hunting was intensified by the joy of possessing. Whatever an ardent young collector may be collecting for the time being is very precious. To be the actual owner of an object he and others have been looking for, is to be wealthy.

There is also a peculiar pleasure in being taught what you want to learn. How grateful a man feels then!

The lads learned a great deal from the above-mentioned chapter—how to make and use egg-drills for boring the sides of eggs intended for the cabinet; how to discharge the contents of the eggs by means of the blowpipe; and how it was best to use a glass bulb-tube for sucking out the contents of the more delicate eggs. Furthermore, they were instructed how to arrange the eggs in the cabinet, and what sort of drawers to make or get made. At their school there was a carpenter's shop, and every boy was a bit of a carpenter—as indeed every boy should be, for there is no handicraft knowledge more useful. So you had only to tell lads like our young friends what to make, and they would certainly manage to make it.

Mr. Southwell recommends the plan adopted

by Mr. Salvin, the distinguished ornithologist, for

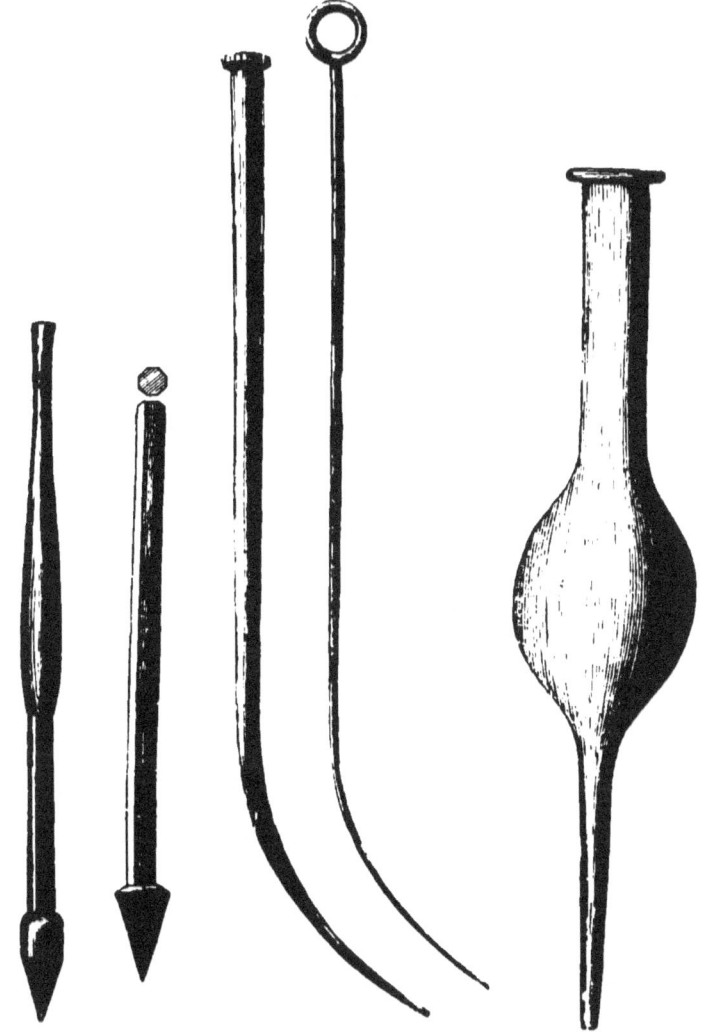

Fig. 29.—Egg-drills.   Fig. 30.—Blowpipe   Fig. 31.—Glass tube
                          and wire.   for sucking eggs.

arranging birds' eggs. Each drawer in the egg-cabinet is divided longitudinally by thin slips of wood into three or more parts, about four to six inches across, as may be convenient. A number of sliding stages are then constructed of *cardboard*, by cutting the card half through at exactly the width of the partition, and bending the sides down

Fig. 32.—Cabinet drawer for eggs.

to raise the stage to the required height. A number of oval holes are then cut by hand, and a thin layer of cotton wool gummed on the upper surface of the stage, the holes being suitable to the sizes of the eggs they are intended to receive.

Between these stages sliding partitions are placed, made of just sufficient height that the horizontal part may fit closely on the wool. These partitions are made of thin wood for the upright part, along which a horizontal strip of cardboard is fastened with glue: on the latter is placed a label bearing the name of the egg displayed on the stage. By this plan the eggs are well shown, and not likely to fall out of their places. Each drawer is then covered with a sheet of glass, to exclude the dust.

CHAPTER IV.

NIMRODS AMONG THE LEPIDOPTERA.

THOSE happy days when the summer is young, and we are as young as the summer! When the summer-tide and one's life-tide are both flowing! Happier days still, when the love of Nature has got hold of a youth who, although

> "Daily further from the East must travel,
> Still is Nature's priest, and by the vision splendid
> Is on his way attended!"

The months of May and June were a charmed season to our enthusiastic lads. Every hour they could get away they passed in the woods, lanes, and fields, or else wandered over commons and through miry swamps and marshes. They made collections of all the birds' eggs they could, and studied the nests of the commoner kinds—how they were built, and the variations in their structure. The orders of Greek architecture do not vary more than the architectural differences in birds' nests.

Then the lush foliage and greenery of May and June made bird's-nest hunting all the more delightful, because it was more difficult than when there was less greenery about, and when any idiot could find a nest. One of their chief pastimes was that of watching birds to their nests. It requires a good deal of patience and keen observation, but it rewards us in the absolute knowledge one gains by getting a good practical knowledge of the characters of birds. It is a double watching—we watch the birds, and they watch us. The urgent affairs of their nests tempt them to all sorts of tricks and artifices to deceive us. But at last the bird takes good heart, and trusts that all will be right. Then it drops into its nest, or enters its hole; and the young watcher conquers by finding the nest, but generously forbears to meddle with it, the delight of having overcome the cunning of the bird and of finding the nest being victory enough—at least, to any boy but a cad.

Then the butterflies had come out. Every boy is a born butterfly-hunter. He cannot resist attempting to capture one of these fluttering, animated bits of colour. It is like owning a morsel of summer. So away go all sorts of lads after them, with caps and jackets generally. Jackets are magnificent hunting-tools with lads. Along comes a butterfly, leisurely stopping at one flower after another, and off goes the boy's jacket. Away he runs after the prey. The sluggish butterfly enjoys being hunted with a boy's jacket quite as much as real red-coated huntsmen say the fox does. But the butterfly generally escapes the jacket.

How could our young friends help being tempted aside after the butterflies, or, for the matter of that, after the moths which came out as they returned to school in the gloaming? As a rule, they were too much interested in other things in the daytime, but in the evening the moths were a sad and almost sole temptation.

Of course they were aware that, whilst butterflies are *diurnal* insects, moths are usually *nocturnal*. They also knew—for the professor had already pointed the fact out—that one could always tell a butterfly from a moth by the antennæ of the former being *clubbed* at their ends. He had also drawn their attention to the law of *mimicry* discovered by that prince of entomologists, Mr. Henry Bates. By this law, insects possessing stings or other self-defences, are often marvellously imitated by other insects belonging to orders as far apart from them as the poles.

Now, if you and I pretend to be what we are not, our conduct is called *humbugging*. If we push it to extremes of deception, it may be denominated lying, and we may undergo the risk of

Fig. 33.—Narrow-bordered clear-wing.

imprisonment for living under "false pretences." But when insects, plants, etc., adopt this policy of duplicity, naturalists call it *mimicry*. There is

Fig. 34.—Hornet clear-wing.   Fig. 35.—Currant clear-wing.

something in a name, after all, despite the remark of Shakespeare.

Look at those British insects called the "clear-wings," for example. They fly in the daytime, and look to all the world like so many hornets, wasps, bumble-bees, ichneumons, and other insects people don't care to make too familiar an aquaintance with, on account of their stings. They not

Fig. 36.—Broad-bordered clear-wing.

only marvellously resemble sting-possessing insects, but some of them fly in the same manner. You don't see them fluttering like leaves, after the

Fig. 37.—Humming-bird hawk moth.

manner of butterflies, zigzagging about the road; they fly straight, and hover over flowers. One nearly allied species to these clear-wings, the humming-bird hawk moth (*Macroglossa stellatarum*), has the under-wings somewhat resembling the above-mentioned insects. If you rub off the "dust," as it is called, from the wings of butterflies and moths, you make them "clear-wings"—that is, you

Fig. 38.—Transparent burnet.

Fig. 39.—Red-belted clear-wing.

get at the transparent membranes over which the "scales" are arranged like tiles on a housetop.

Every species of butterfly and moth has a specific set of scales sometimes two or three patterns in

Fig. 40.—Scale from *Vanessa urticæ* (magnified).

different parts of the wings. Most of them are very beautiful objects when seen under a microscope. The little grains of dust which come off the wings of a captured butterfly so readily, then expand into distinct shapes with characteristic markings. Here are the magnified scales of the smaller tortoiseshell butterfly (*Vanessa urticæ*) and the white cabbage butterfly (*Pieris brassicæ*), etc., for instance, showing the lines or rows of small

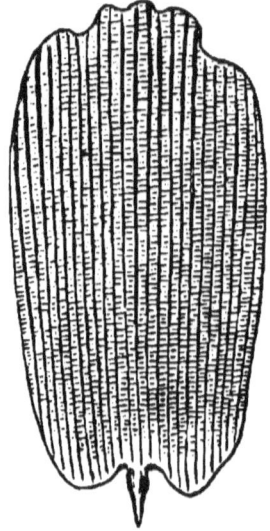

Fig. 41.—Scale from *Pieris brassicæ*. Fig. 42.—Battledore scale from *Polyommatus alexis*.

dots, which are so close together that they look like lines.

The more highly you magnify these natural objects, the more beauties do you discover.

It is the same with the *eggs* of insects. Few people are aware of the lovely, ivory-carved-like appearance of the eggs of the bluebottle, or even

NIMRODS AMONG THE LEPIDOPTERA.   57

of the common house-fly. The eggs of butterflies and moths, like their scales, are often exquisitely adorned with dots, lines, tubercles, and stipples, which require a good magnifying power properly

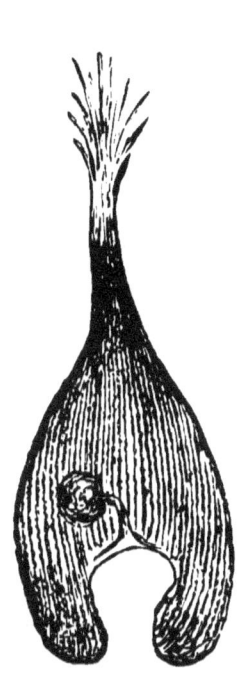 

Fig. 43.—Scale from *Pieris brassicæ*.

Fig. 44.—Scale from *Hipparchia janira*.

to behold them. Pick up the eggs of any butterfly or moth (and I don't see why butterflies' eggs should not be as well known as birds' eggs; they are certainly quite as beautiful, if they do not excel them in beauty). It would be a capital thing for one or two young fellows like our friends to begin collecting insects' eggs.

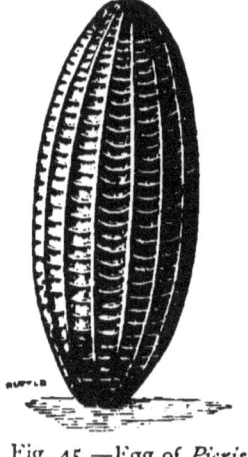

Fig. 45.—Egg of *Picris brassicæ*.

Fig. 46.—Egg of *Vanessa atalanta*.

Every young naturalist ought to be something of an artist. Drawing ought to be as much resorted

Fig. 47.—Egg of the buff-tip.

Fig. 48.—Egg of meadow brown butterfly.

to for the expression of certain ideas as writing

A naturalist ought to be able to sketch what he describes. "Word-pictures" are all very well, but real pictorial representations are better in natural science.

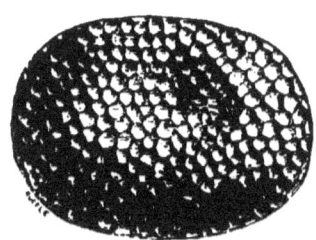

Fig. 49.—Egg of the common magpie moth.

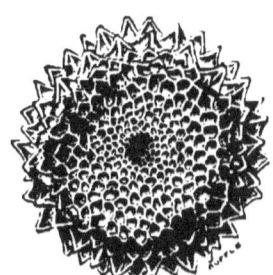

Fig. 50.—Egg of *Polyommatus corydon*.

These highly magnified pictures of the eggs of some of our commoner species of butterflies and moths will give a good idea of what I mean. My

Fig. 51.—Egg of the cabbage moth.

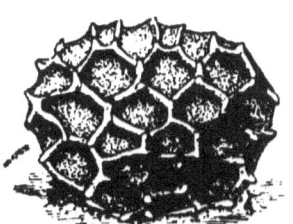

Fig. 52.—Egg of the small copper.

readers will observe at a glance that each kind is distinguished by a special egg-pattern, just as birds' eggs differ in colour and marking.

These matters had hardly as yet possessed the minds of our young friends; but the ardent love of nature possessed by their science-teacher had caused him to hint at them, and suggest many of these thoughts to the boys of his class. As the summer drew on, he went out on the half-holidays with those young fellows whom he had mentally inoculated with his own tastes.

What glorious collecting and hunting times those were! Nothing was left unnoticed, or uncollected. Every commonplace plant, insect, bird, stone, fungus, moss, became a prize.

You would have imagined, from the personal pleasure the young professor manifested when some lad brought him one of these things, that it was the very rarest or the most instructive object that could possibly have been brought to him. How happy was the ignorant young lad who had the good fortune to find it, and how zealously he looked about to find something else!

Even if the boys had forgotten every object whose name and character they learned, they would have been great gainers; they had learned to observe—to use their own eyes. Many people think they do, when in reality they are using other folk's, and are simply finding what they are told to look for, and nothing else. That is not the way for knowledge to increase.

What hosts of things there are to be observed in our green lanes! what hieroglyphics to be deciphered! The leaves of the bramble, hawthorn, oak, and other shrubs and trees, are marked with zigzag or sinuous markings. They are about the commonest objects to be met with,

Fig. 53.—Mined oak-leaf.

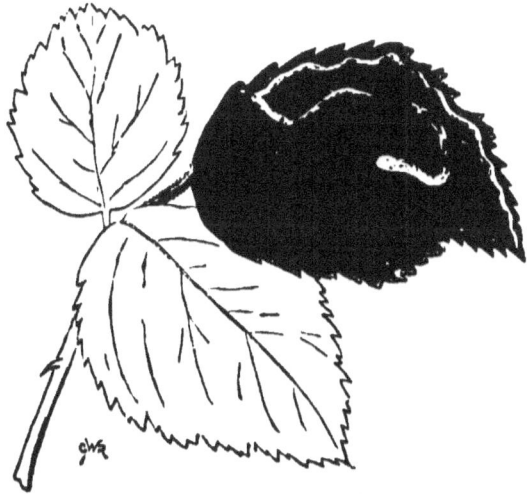

Fig. 54.—Mined bramble-leaf.

and we may therefore be sure they did not escape the professor's ardent class.

They learned that certain very small grubs, the larvæ of one of the smallest moths, in order to be secure, mined beneath the upper and the under skins or surfaces of these leaves, thus getting both food and protection. Holly leaves are favourite ones for another species of moth. If the leaf is still green, probably you will find the minute

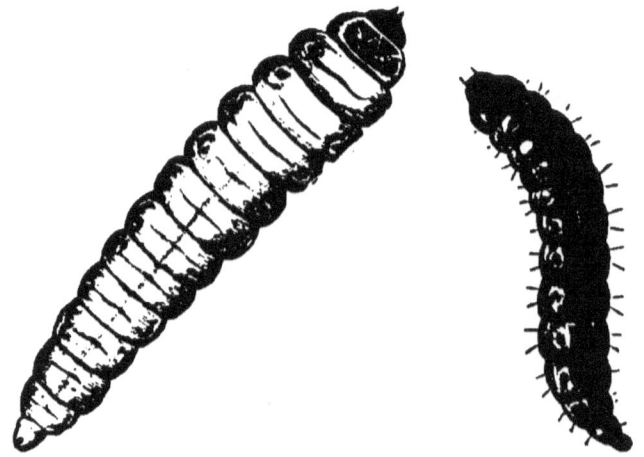

Fig. 55.—Larva of *Micropteryx subpurpurella* (enlarged), the caterpillar which mines the oak-leaf.

Fig. 56.—Larva of *Nepticula durella*, the miner of the bramble-leaf (magnified).

caterpillar housed inside. Then you can magnify it, and take as many observations of it under the microscope as you like.

Many of the boys had rigged up butterfly-nets out of a walking-stick, with a bag mounted as you see beneath; others had the usual regulation equipment. But the professor was not so anxious to collect as to observe, although he well knew that if the boys did not capture something, they would lose interest.

For instance, on their leisurely ramble they came across a tree whose trunk had been bored and drilled very recently, as could be told by the little pile of sawdust at the base. He pointed out that this was the work of that big fat grub which,

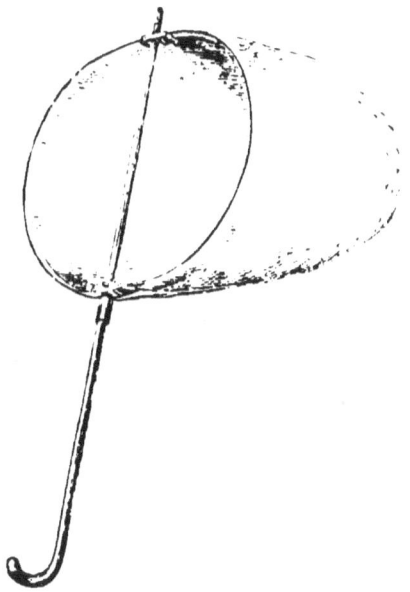

Fig. 57.—Umbrella net.

Fig. 58.—Larva of goat-moth.

it is said, the ancient Romans regarded as a *bonne*

bouche, the caterpillar of the goat-moth (*Cossus ligniperda*). They were not long before they dislodged the creature. It was dropped alive into a larva-box, for further observation in the insectarium, where it was subsequently seen to cover itself with a layer of

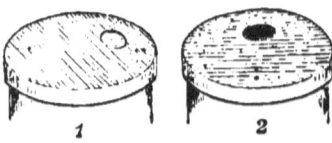
Fig. 59.—Boxes for larvæ.

Fig. 60.—Pupa of goat-moth.

sawdust as a cocoon, beneath which it underwent

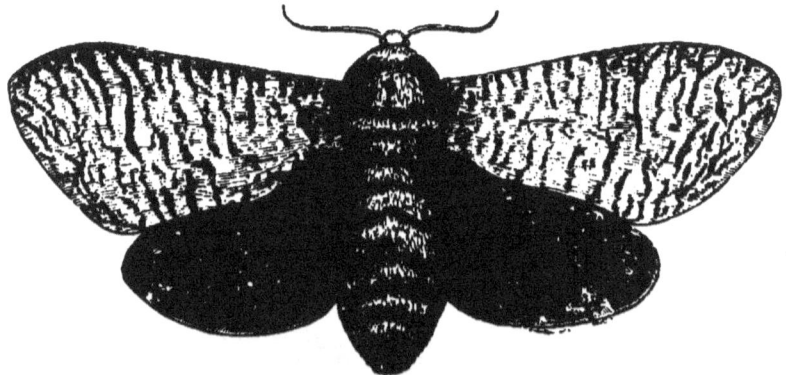
Fig. 61.—The goat-moth.

that marvellous physiological transformation which

changes both interior and exterior of a grub into a butterfly or moth. The beginning of this change, however, had already commenced when the caterpillar entered the pupal state. In that apparently resting stage, the materials elaborated by the greedy and fast-growing grub are worked up into additional tissues, muscles, and organs of locomotion.

The manner in which the various members of the Lepidoptera instinctively prepare for these several changes was pointed out some days afterwards, in the case of the white cabbage butterfly (*Pieris brassicæ*). When it is commencing to form the silk cord which will support the future chrysalis, it bends back its head to the fourth segment of the body, and then turns its head downwards

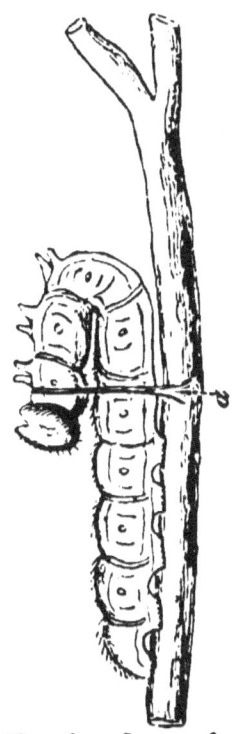

Fig. 62.—Stage of metamorphosis of *Pieris brassicæ*.

on the right side, so as to bring its mouth to the point marked *a* (Fig. 62). There it fixes its first line of silk; then it carries its head over to the left side, spinning a silk line all the time, and after-

F

wards fastening it down on the left side—again bringing the silk line to the right side, and fastening it down. The caterpillar repeats this process about forty times, until as many silk lines have

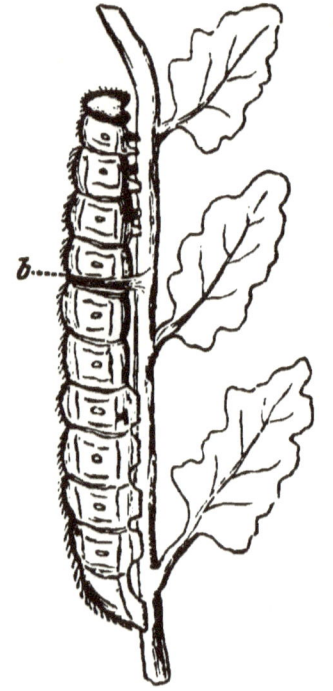

Fig. 63.—Another stage in metamorphosis of *Pieris brassicæ*.

Fig. 64.—Chrysalis of *P. brassicæ*.

bound it to the spot it has elected to pass the chrysalis stage in. The silk is drawn so tightly that the creature appears as if its body would be severed in twain. Then comes the task of releasing the head from this bent-back and tied-down condition; but the caterpillar takes advantage of the elasticity of the freshly made silk, and is then in the position shown at *b* (Fig. 63), where it rests until the chrysalis is formed. Then the old caterpillar skin is thrown off from under the silken cords, and the true chrysalis is seen (Fig. 64).

The professor talked a good deal about cater-

Fig. 65.—The emperor moth.

pillars and chrysalids. It is astonishing how much can be said about them. He showed that the hairy kinds are never eaten by birds, and that green caterpillars were seldom hairy. This was because their green colour screened and protected them by causing them to resemble the foliage on which they feed. Even fish will not eat the hairy caterpillars—only the green or naked kinds. Consequently the hairy species rejoice. They seem to know they are safe. Many of them can roll themselves into a ball, like that of *Arctia caja*, and then they are about as edible as a live hedgehog would be to a hungry man! These hairy caterpillars are often brilliantly coloured and prominently marked, as if they invited notice instead of concealment; and as if they knew they could say to the birds, " Here we are ; eat us if you dare ! "

Fig. 66.—Caterpillar of emperor moth.

Some of the commonest of these protected caterpillars are those of the little eggar moth (*Eriogaster lanestris*). Indeed, this species is doubly protected, for not only are the bodies of the caterpillars more or less hairy, but they have the power of spinning a strong silken tent or web, beneath which they safely consume the leaves of the plants on which they feed. They live in social colonies, and this habit must be of great service to them. In dry, hot summers, the hedgerows are not unfrequently festooned with the webs for yards in length, and the leaves are stripped off as if a fire had passed over them. Then as to the *origin* and *meaning* of many of the dots and markings on caterpillars. Look at those on the Sphinx family, for instance. The caterpillars moult a good many

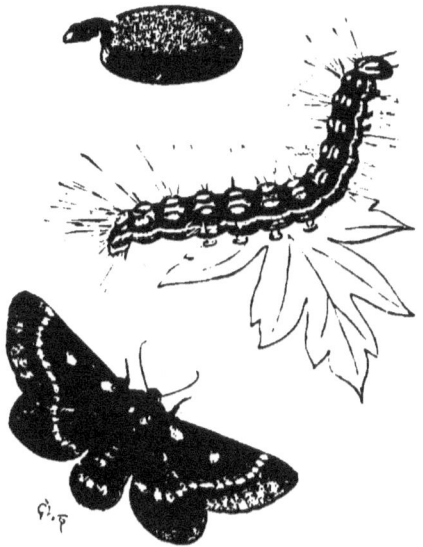

Fig. 67.—Caterpillar, cocoon, and imago of small eggar moth (*Eriogaster lanestris*).

times in the process of growth. Every time the body is differently marked, although it is the last set of markings which are usually recognized as distinctive. The caterpillars of different species of Sphinx moths are known by their different markings; and the professor showed his attentive students how, in perhaps the most abundantly and prominently marked species, the markings on each moult more or less resemble those of each of the various species of the group. In other words, that a single caterpillar repeats, in the development of its own individual life-history, the changes which have caused the development of species.

The manner in which caterpillars fed furnished

Fig. 68.—Bedstraw hawk moth.

another theme for description and discussion. It arose one day from one of the boys being almost startled at finding that buffoon of the caterpillar world, the larva of the puss moth (Fig. 69).. There it was, sitting like a dowager, on the edge of a leaf, comfortably cutting and coming again as its voracious appetite dictated, and pretending to look fierce, like a Japanese warrior. Now, pretence is like *advertising*—it's no good unless you do it

Fig. 69.—Caterpillar of puss moth (*D. vinula*).

well, and be earnest in doing it. So the caterpillar of Vinula has succeeded, as its portrait shows.

The resemblances borne by many of the *micro-*Lepidoptera, that is, the *minute* moths (gnats, ignorant people call them), to other objects, for protective purposes—such, for instance, as their resemblance even to the droppings of birds on leaves; their green, brown, and lichen tints, all more or less protective—were discussed. These

Fig. 70.—Vapourer moth, male.   Fig. 71.—Ditto, female.

discussions gave quite a zest to the discovery of specimens confirmatory of the theory.

Then it was shown how the females of certain moths were *wingless*—how, in one species, they practically never advanced beyond the caterpillar stage; in another, hardly beyond the pupa stage, and so on; how, for protective purposes, one wingless female had six long legs, and resembled a spider so much that you would hardly have known the difference without counting the legs first, which is a thing few birds do. In all these species the male is fully winged as usual.

As the summer drew on, and the holidays approached, you may be sure that not only our two friends, but nearly the whole class, determined to indulge to the full in the newly discovered pleasure of observing and collecting. They were put up to all kinds of dodges—how to proceed, what to look for, how to preserve it, etc. One lad had a fad for beetles, another for shells, and several of them for anything and everything they could get. Butterfly and moth collecting, however, are nearly always the first subjects boys take to who have a natural-history turn of mind. Their teacher was perfectly aware of this, and therefore encouraged them. He knew that many such collectors would proceed to other studies, and would collect other objects; but he was aware that the habit engendered by butterfly and moth hunting would abide in any other pursuit.

So he initiated them into the mysteries of beating and hunting willows, brambles, heather, etc., after dark, with a lantern and net; also in "sugaring"—which is about as interesting a pursuit as a romantic lad could be introduced to. The results of sugaring can only be known after dark, and the young collector feels something brigandish or poacher-like as he goes about with his dark lantern, examining the tree-trunks which have been smeared with a beery mixture of sugar and treacle and a little rum, etc. The smearing is generally done just before dusk, and the baited spots are visited when darkness has set in. Then it is astonishing what a number of guests have invited themselves to the spread. The only circular issued to them was the smell of rum,

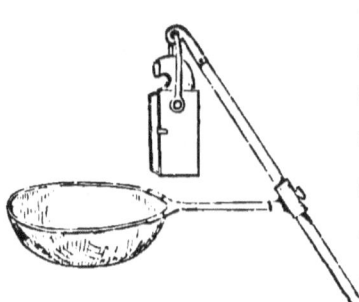

Fig. 72.--- Lantern and net.

Fig. 73.—*Diphthera Orion.*

Fig. 74.—Small black arches.

Fig. 75.—*Hyria auroraria.*

or aniseed, or whatever else had been put in the sugaring mixture. Insects have an almost phenomenal development of the sense of smell. It is so keen that if you carry some species of imprisoned virgin females you have reared yourself (although, perhaps, of a rarish kind) in a perforated

Fig. 76.—Clouded buff moth (female).

Fig. 77.—*Selidosema plumaria* (male).

box, all the male insects for miles round will come trooping to her, like so many mediæval brave knights serenading an imprisoned damsel! Our own sense of smell, although it beats the spectroscope for keenness

Fig. 78.—*Lithosia quadra.*

of detection, is dull and sluggish when compared with that possessed by many insects.

Fig. 79.—Scarlet tiger moth.

We ought to be thankful such is the case. Flowers would never nave possessed perfumes if

Fig. 80.—Early thorn moth.

flower-hunting insects had not been gifted with a keen sense of smell. Perhaps that sense and the perfumes have originated side by side, and helped to develop each other.

Fig. 81.—*Thyatira batis.* Fig. 82.—Male and female of oak hook-tip moth.

Those charming sultry summer evenings with

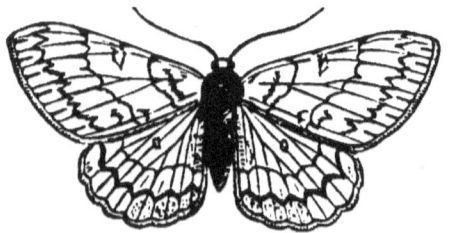

Fig. 83.—Pale oak beauty.

the sugaring-pot, the collecting-net, and the dark lantern! They leave too delightfully and en-

Fig. 84.—The large emerald moth.

duringly keen a sense of pleasure for all the years

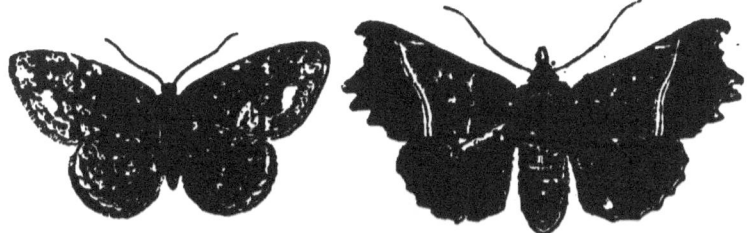

Fig. 85.—The brindle white-shot moth.   Fig. 86.—The herald moth.

of after-memory to be able even to obliterate them.

Fig. 87.—Female of four-spotted footman.

Of course, a large number of the butterflies

caught in the daytime, and of moths captured at night, were common species. But that is nothing

Fig. 88.—Male of four-spotted footman.

Fig. 89.—*Philophora plumigera.*

at first, to the young collector, to whom the joy of possession counts for a good deal.

Fig. 90.—Fox moth.

Fig. 91.—Cream-spotted tiger moth.

But as our friends varied their evening walks,

sometimes in the woods, at others in the green lanes, or over the heath, or down by the marsh,

Fig. 92.—The gold-shot moth.   Fig. 93.—The black-vein moth.

they found that different species of moths were peculiar to these various habitats, or localities.

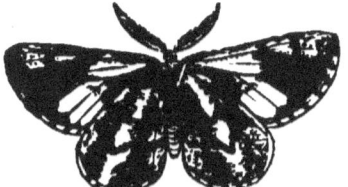 

Fig. 94.—The bordered white moth.   Fig. 95.—*Selidosema plumaria* (female).

Perhaps it was because each place is so physically different, and therefore different flowering plants

Fig. 96.—The chalk carpet-moth.   Fig. 97.—The small emerald moth.

grow in each. Moreover they soon learned that *soils* and rocks regulated the distribution of species.

You find one kind only on light lands, another only on heavy. Species are only met with peculiar to limey or chalky strata, and others where sandstone, shale, or other rocks prevail. It is,

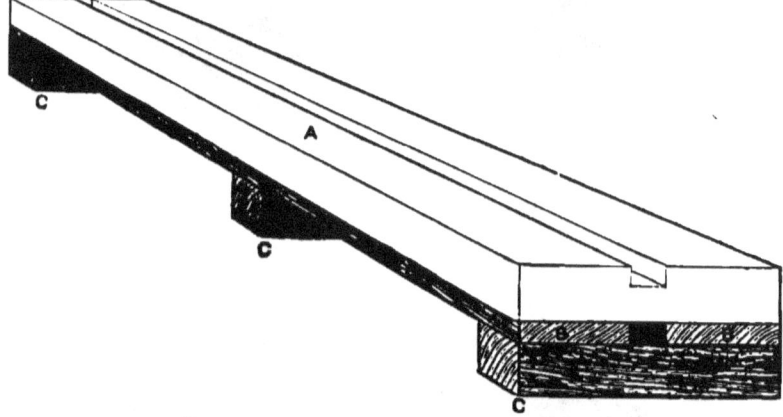

Fig. 98.—Setting-board for Lepidoptera.

perhaps, this wonderful physical and geological differentiation of the earth's terrestrial surface

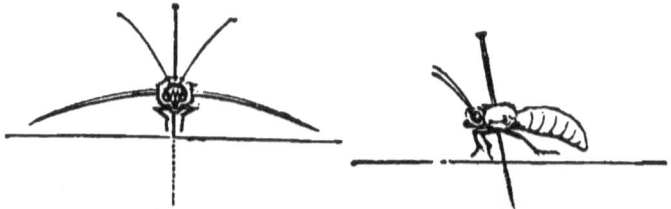

Fig. 99.—Front view of properly pinned-out insect.   Fig. 100.—Side view of properly pinned-out insect.

which has largely assisted in developing species of flowering plants, and, through them, of many kinds of insects.

The use of the cyanide-bottle for instantly killing specimens, and how to set them out and properly strap and pin them down afterwards,

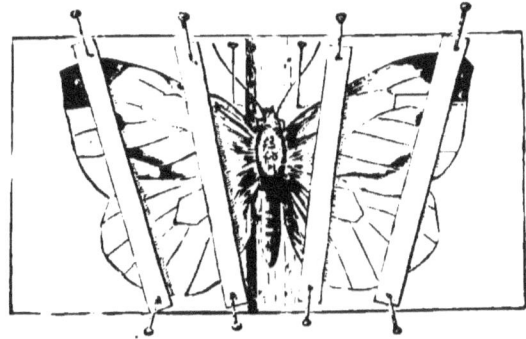

Fig. 101.—Mode of setting out Lepidoptera on level board.

were all carefully explained to the young naturalists. There is no part of any of these mechanical apparatuses which any ingenious youth cannot make or rig up for himself.

Fig. 102.—Moth set out on cork saddle.

In addition to the long, flat, grooved setting-board above shown, a grooved cork saddle is frequently used, and the accompanying side and

G

front views of the insects when pinned down, will give a clear idea of how they are arranged. The use of thin but stiff paper straps for holding down the wings, antennæ, etc., and arranging them in the freshly set-out butterflies or moths until they

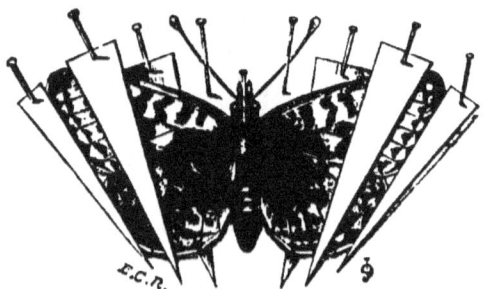

Fig. 103.—Example of four-strap setting.

have assumed the rigidity desired in the cabinet, will also be made evident by our illustrations.

At any rate, our professor did his best to start the lads, to whom he was much attached, to observe, collect, and arrange for themselves. If they failed to take advantage of his experience and ready help, it was their own fault.

CHAPTER V.

HOLIDAY RAMBLES AND ADVENTURES.

MUGBY SCHOOL broke up for the holidays earlier in the summer than others. Jack Hampson went home with a limited but enthusiastic stock of knowledge concerning common natural history objects. If he did not know much, at any rate he had learned to make a country walk more enjoyable than he had thought such perambulations could turn out. He was not long in displaying his newly obtained knowledge; and even if he were a little proud of it, and rather paraded it a trifle, it was pardonable. But he was not a prig, so there was little of either brag or show in his ready display of what so much interested him. Rather, it was the zeal of a proselyte. He wanted others to enjoy his own new-born pleasure. He could not keep it to himself; it bubbled over irresistibly.

Now, that is the sort of human being—lad or man—to make converts! You cannot quarrel with him. You may pretend to laugh at him, or poke fun at his notions; but, if he has any "go," it is ten to one the enthusiast will convert you.

The first convert Jack made was his eldest sister—a strong, active lassie between eleven and twelve years old. His younger brother was also bitten, but not so rabidly, for younger brothers don't like their elder brothers to see they can do what they like with them, and cram any notions they please down their throats. Then there were a couple of cousins, fine lads, from another school, who presently joined the Hampson party, and they fell victims to the mania for collecting and preserving. The time for birds'-nesting was, unfortunately, nearly over; but moths and butterflies were to be had for the hunting, and the delights of chasing them in the daytime and of sugaring for them in the evening, were duly indulged in.

I have said that the young professor knew that an energetic boy like Jack would soon extend his observations further afield; that presently he would have captured nearly all the common species of Lepidoptera in his neighbourhood, and would be sighing for something else to conquer. So he told him to collect anything he saw in his rambles—anything, he said, except tombstones!

One cannot take a walk a hundred yards in the country without seeing plenty of natural objects of which we are in perfect ignorance. We don't know even their names, to say nothing about the structures, life-histories, and general habits. Nowadays, a grown-up man or woman is ashamed of being unable to read. Yet how many millions of people are not ashamed of being unable to read this great Book of Nature, written within and without like the prophet's scroll, by the finger of the Almighty Father Himself?

Jack was to send all such general objects as he was not acquainted with to the teacher, who had promised to name them. Willie and he were to write to each other, and duly report progress as to their several finds. To add to their zeal (should there be any danger of its flagging), Jack's uncle had promised him what he now desired to possess more than anything in the whole world, a student's microscope. His birthday was only three weeks off, and the present was expected to crown that auspicious occasion. As a matter of fact, it did so.

The young students begged and procured a room over the stables, where they could keep their treasures without littering up the house, or frightening the servant-maids with their nasty things. That was a grand room on wet days, especially after the precious microscope had arrived.

Several packages of unknown odds and ends, chiefly insects other than butterflies and moths, had been sent to the professor, who seldom lost much time in telling them what they were—for, after all, they were among the common objects; few rare ones appeared.

Fig. 104.—Gall insect (*Cynips kollari*), nat. size and enlarged.

Among these were the numerous galls on plants, shrubs, and trees of all kinds, made by certain kinds of insects, so as not only to conceal their young from enemies, but place them in the midst of plenty of food.

Some of these gall insects attack the undeveloped leaf-buds of oaks, preventing the leaves developing, or the branch from growing, and causing them to assume the appearance which has given them the name of artichoke galls. Then there are the various oak-apples, button-galls, oak-spangles, some of which are often thought by young naturalists to be a kind of parasitic fungus. They occur usually on the under-surfaces of oak-leaves, as the illustrations indicate. The minute dipterous insects whose venom and irritation set up the vegetable inflammations which

Fig. 105.—Insect of the button gall (enlarged).

Fig. 106.—Gall insect (nat. size and enlarged).

result in these curious growths, were sketched, both natural size and enlarged. Galls were found in many flowering plants, such as the seed-vessels of the pretty germander speedwell. The lovely "robin-redbreast's cushion," on the branches of the wild roses, is one of the best known. So are the oval, reddish, wart-like lumps on the leaves of willows, and the swellings on the stems of the ragwort. They bulge out certain of the seed-vessels of umbelliferous plants, and cause the thread-like leaves of the common yarrow to develop into vase-like cups. The stem of the thistle often expands into large oval shapes, and if you cut one open, you find it divided into compartments, in each of which is lodged a fat grub. The upper surface of the leaves of the ground ivy are often covered with little hairy galls, in each of

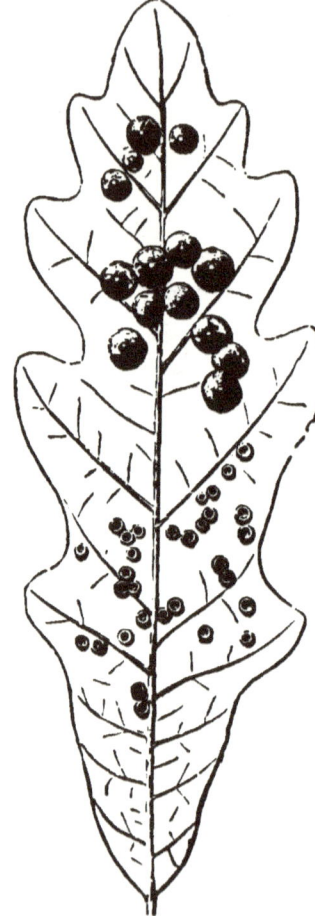

Fig. 107.—Galls on oak-leaf. The upper portion is crowded with galls called "spangles;" the lower, with "button galls."

which is the larva of another gall-fly, or *Cecidomya*. Even the stings of the nettle do not debar another species from making galls, both on leaf, flower-stalk, and leaf-stalk. You find them in abundance on the ends of elm twigs, as well as on the leaves; on the birch leaves, one species occurs on the upper surface, and another on the lower. The oak is the favourite tree for these insects; more than thirty different species make galls on it. Perhaps the most noteworthy are the hard, conical

Fig. 108.—Section of "spangle" gall (magnified).

Fig. 109.—Section of "button gall" (magnified).

barnacle galls, which may be found clustering the smaller branches of the oak. The willow is another favourite tree for them. In addition to the oval kinds found on its leaves, you may discover another which clusters along the edges. Two species of galls are not uncommon on poplar leaf-stalks.

The boys noticed round pieces cut out of the leaves of the rose-trees in the garden; and one day they caught the offender right in the act. They watched the creature—the leaf-cutter bee (*Megachile Willoughbii*)—turning round on its own body as a pivot, cutting the leaf as it rotated; and they could not help admiring the ingenious way with which it flew off, just when they imagined the bee would come to the same fate as the silly woodman who sat astride the big bough he was sawing off. But the bee was not such a fool. It flew away with the round bit of leaf just at the precise

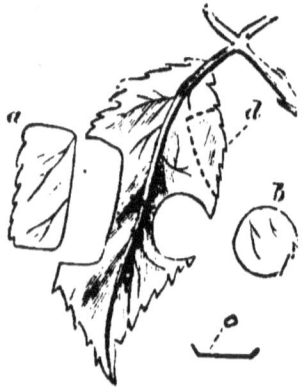

Fig. 110.—Rose-leaf cut by leaf-cutter bee.

Fig. 111.—Leaf-cutter bee (*Megachile Willoughbii*) cutting a piece of leaf for its cell.

moment when it was cut, and used it at once to line its cell with.

The two-winged, or *dipterous*, insects are common enough, but, although some of them are remarkably pretty, adorned with red, blue, and golden metallic tints, the majority are of a dun colour. The *coloured* Diptera are almost in every case flower-visitors. It is an example of the old proverb, that you can tell a man by the company he keeps. So you can a fly, as a rule. The

Fig. 112.—*Conops ralipes*, male.

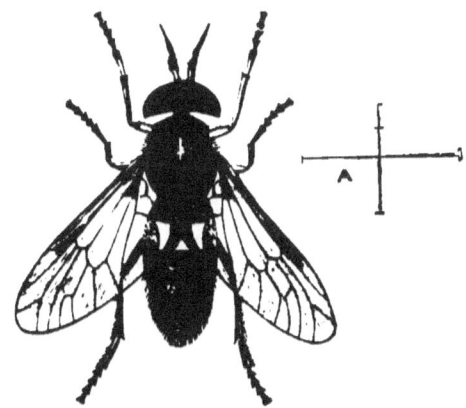

Fig. 113.—Golden-eyed gadfly.

colours of flowers are associated in insects' minds with the pleasure derived from finding their food there. A colour-sense is thus developed. Colours produce pleasant associations. Those insects which

Fig. 114.—*Myopa testacea*, male.

are themselves coloured become all the more acceptable to their mates. Hence the colouration of butterflies, and of fruit-eating and flower-visiting

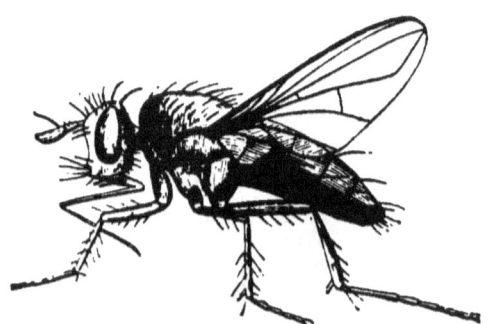

Fig. 115.—*Bucentes geniculatus*.

birds, like the trogons, macaws, parrots, sun-birds, and humming-birds.

Then, again, in such flies as Bombylius we have the same kind of mimicry as in the clear-wing moths. There are several species of this fly, all of which more or less resemble bumble-bees, both in their mode of flight, shape, and even the sounds they make. As you see them flitting from one flower to another, and hovering and creeping about

Fig. 116.—*Asilus Crabroniformis*, female.

them, if you were not an entomologist, you would be certain it was some sort of a bee, and of course had a sting. It has nothing of the kind; it only pretends to have one.

These two-winged, flower-haunting flies are very fond of visiting the numerous species of flowers belonging to the natural orders Compositæ and Umbelliferæ. Indeed, they are among the chief

94 THE PLAYTIME NATURALIST.

agents in cross-fertilizing such flowers. It is there you will probably find the Conops shown in our illustration, trying to make birds believe it is some kind of wasp. One species, *Conops quadri-*

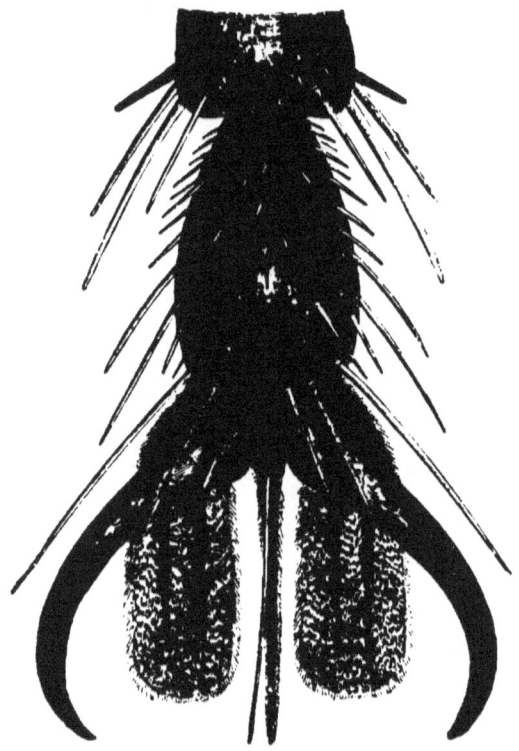

Fig. 117.—Foot of *Asilus Crabroniformis* (magnified).

fasciata, has almost exactly the same colours as a wasp. Some of the species of the Syrphidæ are similarly marked, and actually go by the name of "wasp-flies" on that account.

Myopa testacea suggests the ludicrous, on account of its general gouty appearance. *Bucentes geniculatus* is abundant everywhere during the whole summer and autumn. It is nearly as large as a house-fly. *Asilus crabroniformis* is perhaps the largest and strongest species of British Diptera, the female being larger than the male, as is usual with most insects. The boys had to go to

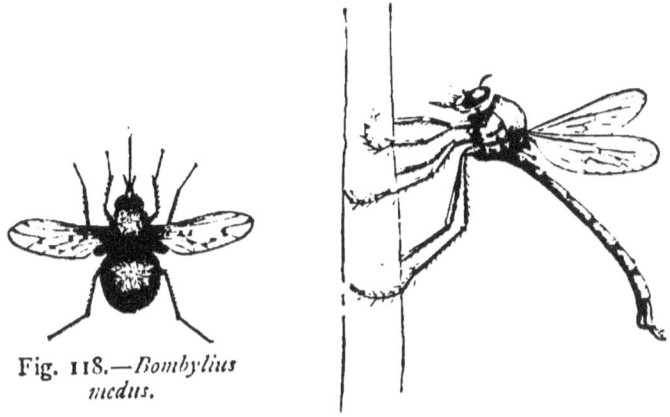

Fig. 118.—*Bombylius medus*.

Fig. 119.—*Leptogaster cylindricus*.

the heaths to find this fly, for it is rather singular in its occurrence. Like many of the others, it adopts mimicry, or "false pretences," if not as a profession, as a protection. Hence its popular name of the "great hornet fly." It is rather a fierce-looking creature, although its colouring is rich, and its bronze-green, compound eyes are lovely objects under the microscope. So also are its feet, with their remarkable pads. On the other hand, quite the opposite of Bombylius, we have *Leptogaster cylindricus*, with a long and slender body, giving it a strong resemblance to the smaller dragon-flies. It is one of the commonest of our Diptera, and may frequently be found clinging like a winged sloth to the stems of plants. Its feet are remarkably adapted to this sloth-like habit.

Fig. 120.—*Anthomyia pluvialis*.

I have given only the pith of the information conveyed in the professor's genial letters to the young collectors, as I thought that was really what my own readers would care about knowing. But the hints and practical knowledge they gleaned therefrom made their pursuits all the more interesting. He also gave them a few ideas of what to look for, and especially what to do on wet days when they couldn't get out, for those are terrible times for boys, and especially to impatient boys.

"For instance," said he, "there is the common house-fly. Now, who knows anything much about

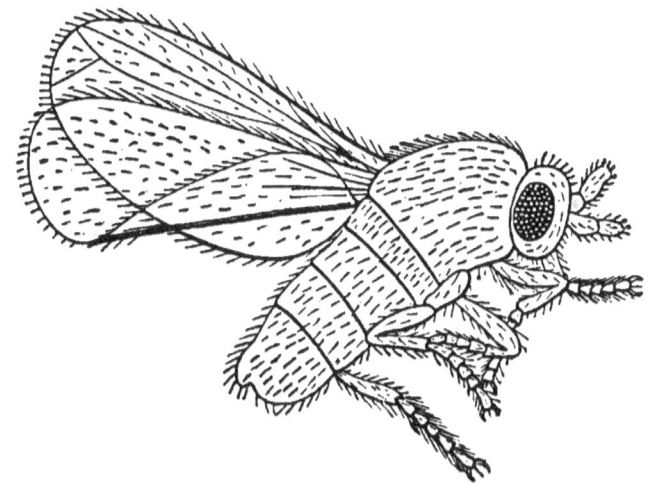

Fig. 121.—The common house-fly (enlarged).

it, except that it's a nuisance? Try and find out all you can about it, its eggs, grub, etc. There isn't one person in a thousand knows anything about these things. They don't know what becomes of the house-flies in the winter; they don't know where they come from in summer They might be freshly created every season, and rubbed out of existence at the end, for what most people who claim to be intelligent know about them or their habits."

Fig. 122.—Egg of house-fly (magnified).

That is perfectly true. Indeed, the boys had never before given a thought to house-flies, although they had given them a good many whacks—or had tried to.

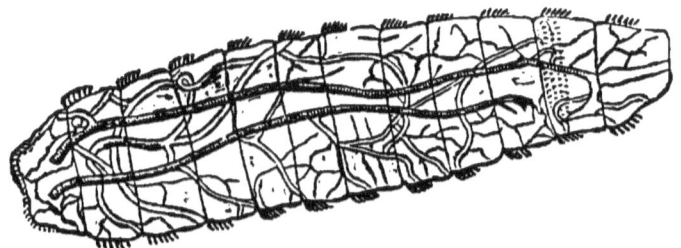

Fig. 123.—Maggot, or larva (magnified), showing tracheal or breathing system.

So now they set to work. A house-fly was soon caught, and examined with a low magnifying power.

A few eggs were found on a cold leg of lamb

which had been left out, very considerately, by the cook. Their diameter was not more than the thirtieth part of an inch. The day after finding them, and imprisoning them (with a bit of meat to keep them company), the eggs were hatched, and the grubs were lively. Beneath the microscope, the nearly transparent skin allowed the air-breathing and circulatory system to appear. The weather was hot, and the meat "high," so the maggot fed well, and proceeded to the chrysalis stage; thence to emerge, in about eight or ten days from finding the egg, to the fully developed house-fly.

Fig. 124.—Chrysalis of house-fly (enlarged).

"Wet Days with the Microscope" would not be a bad title for a book. Nor would a small microscope be a bad companion at the seaside or in the country on such occasions. It would be infinitely better than flattening your noses against the window-panes, and grumbling because it was raining.

So there was always material enough for such inauspicious occasions; and the change of occupation made the sunny days, when collecting was possible, all the more enjoyable. Moreover, Jack had been taught the rudiments of pinning

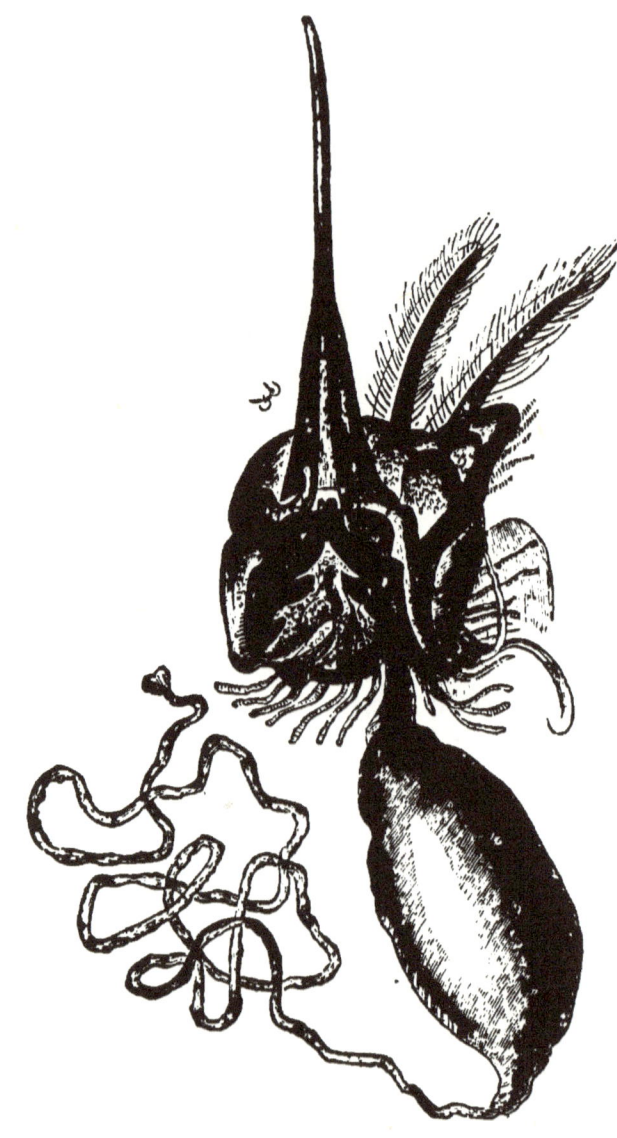

Fig. 125.—Sting, poison-bag, and poison-gland of humble-bee.

down and dissecting, with needles, such objects as worms, caterpillars, etc., and also what to look out for in their anatomy. So the stings of bees, wasps, etc., were dissected out, and mounted on glass slides for microscopic examination; and then all the lookers-on were surprised and delighted at the subtle mechanism they beheld.

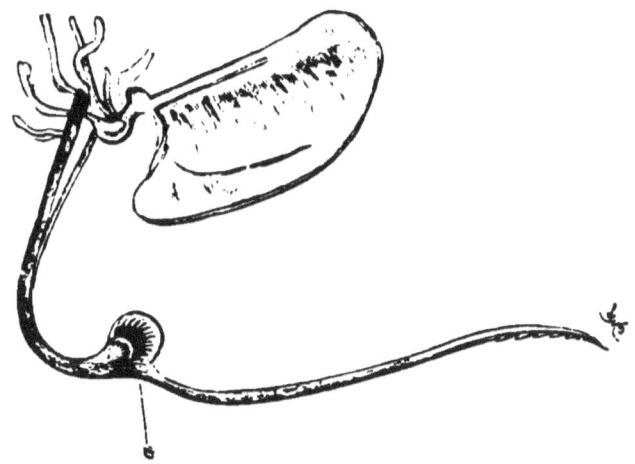

Fig. 126.—Lancet of sting of humble-bee.

That highly domesticated insect, the flea (*Pulex irritans*), came in handy, in lieu of outdoor entomological spoil. The wonderful anatomy of a common caterpillar (one of the easiest objects to dissect) was carefully made out. The compound eyes—*thousands* in number, all clustered together —of a moth, and the proboscis of the same

Fig. 127.—Sting, lancet, and poison-bag of wasp.

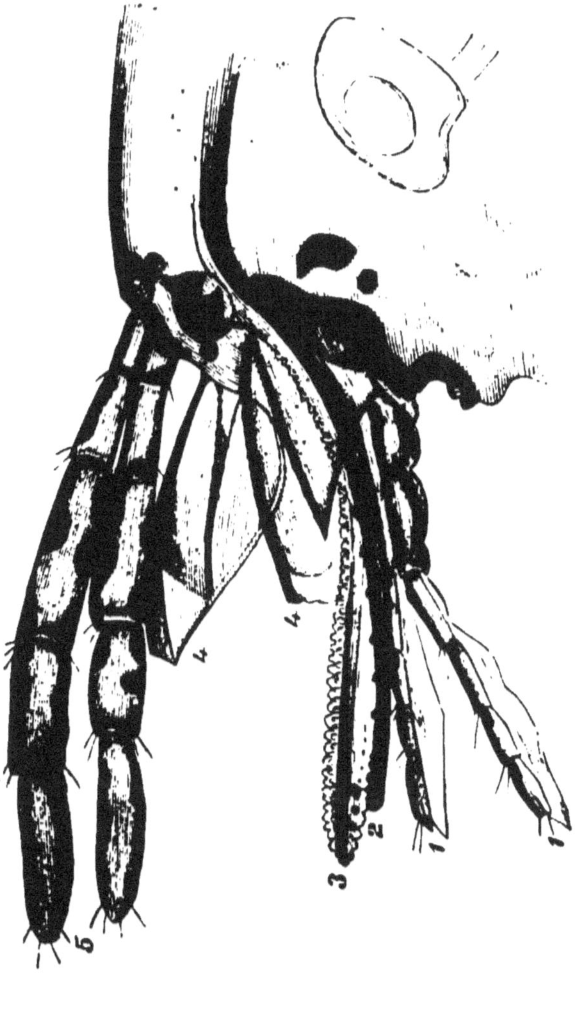

128.—Tongue and lancets of common flea (highly magnified).

Fig. 129.—Lancet of wasp-sting.

creature, were studied, the latter being really two long lips articulated together into a sucking-tube. It was like a natural-history panorama. You could vary the scene as frequently as you liked by putting a different object under the microscope, or even examining a different part of the body of the object. The teeth of the blowfly and house-fly caused much surprise.

Jack soon drilled his brothers, sister, and cousins into his service. They were told to bring to him all the "nasty things" they could find. The lads brought them in abundance, thinking to chaff him and disgust him. The tables were turned when they found that anything and everything was welcome—grubs, flies, worms, thousand-legs, plant-lice, spiders—anything.

"What a rum chap he is!" said his cousins.

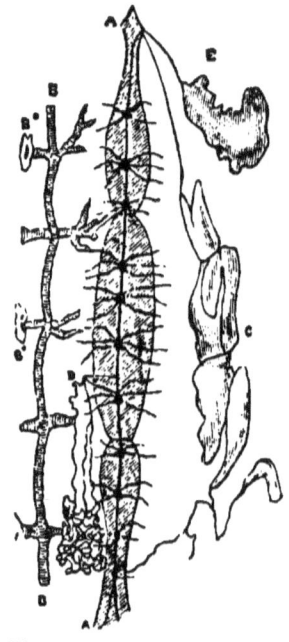

Fig. 130.—Anatomy of a caterpillar. A, Digestive apparatus; B, trachea or breathing-tubes; C, silk-gland; D, liver; E, salivary gland.

But they all learned a great deal about these "nasty" things—how the glow-worm was a wing-

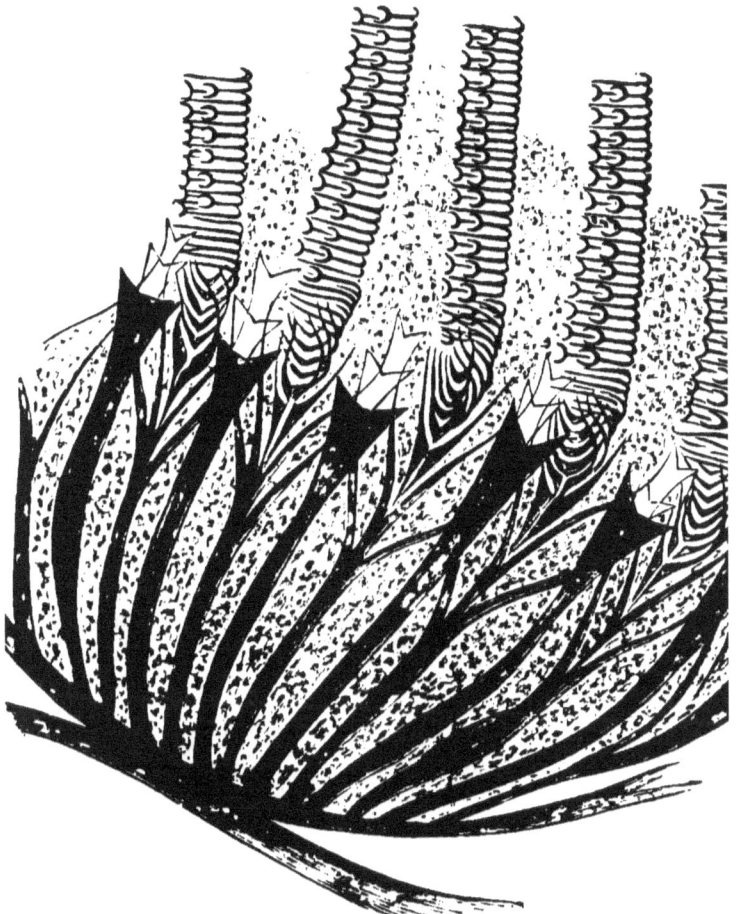

Fig. 131.—Teeth of the blowfly (highly magnified).

less beetle, the male beetle being able to fly; how the plant-lice (aphides) were able to secrete the sweet fluid called "honey-dew," of which ants are so fond that some species actually keep flocks of

Fig. 132.—Head of moth, showing compound eyes, antennæ, and coiled proboscis (magnified).

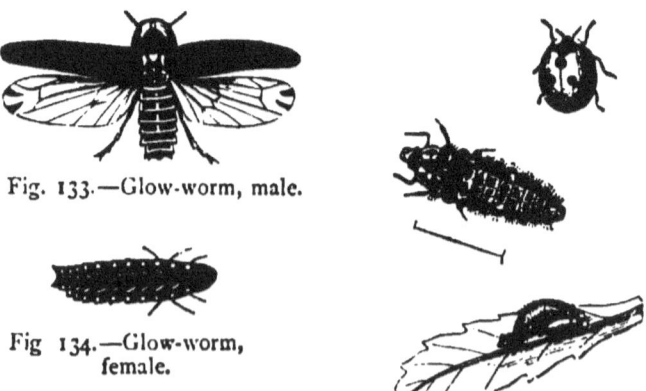

Fig. 133.—Glow-worm, male.

Fig 134.—Glow-worm, female.

Fig. 135.—Lady-bird beetle and its larva and pupa.

aphides as we do milch kine, and regularly milk them! How the lady-birds (Coccinella) are car-

Fig. 136.—Winged aphis (magnified).

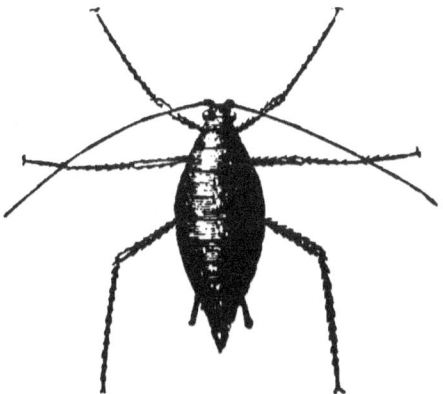

Fig. 137.—Wingless aphis (magnified).

nivorous beetles, and do as much harm to these flocks of aphides as wolves would among a flock of untended sheep; how the larvæ of lady-birds are as unlike their parents as caterpillars are unlike butterflies and moths, etc.

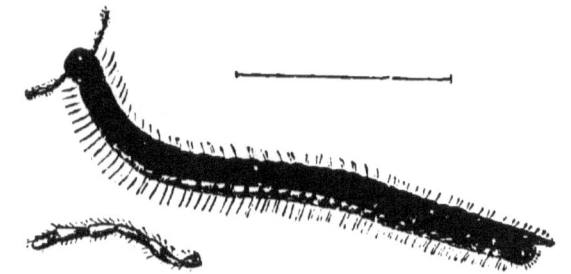

Fig. 138.—"Thousand legs" (*Julus terrestris*).

Spiders, again, were caught and examined. Their eyes, eight in number, big and little, caused much admiration. The big eyes looked like the "cat's eyes" or noble opals set in rings, and their bodies seemed as covered with hair as a leopard's. The spinnerets, by means of which they make those wonderful stranded ropes we call "spider's threads" (some of which are made up of hundreds of twisted lines), were also examined under the microscope.

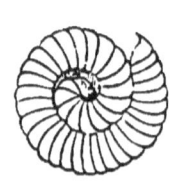

Fig. 139.—Ditto, coiled up.

Talking of spiders reminds one of their nests and homes, as well as their webs. Here is the nest of *Theridion riparium*, for instance, formed of pellets of earth made to hold together, and slung like a hammock from some twig. But for exquisite beauty and admirable suitability for such extra-

Fig. 140.—Track of *Hylonomus fraxini* on the wood beneath the bark of a tree.

Fig. 141.—Imago of *Hylonomus fraxini*.

Fig. 142.—Ditto (magnified).

Fig. 143.—Larva of ditto.

Fig. 144.—General form of main track or channel.

110   THE PLAYTIME NATURALIST.

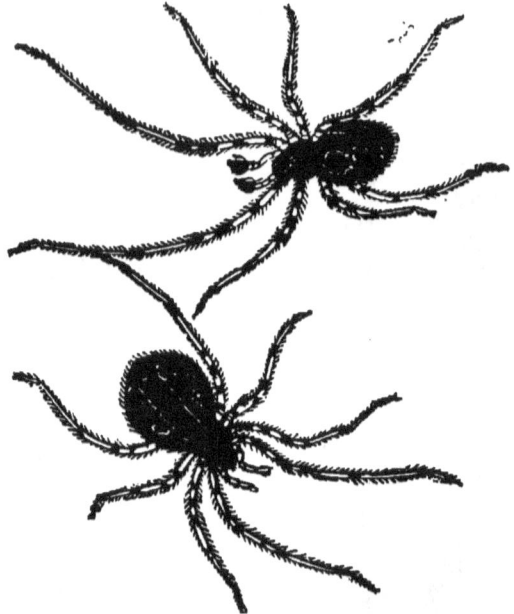

Fig. 145.—*Theridion riparium*, male and female.

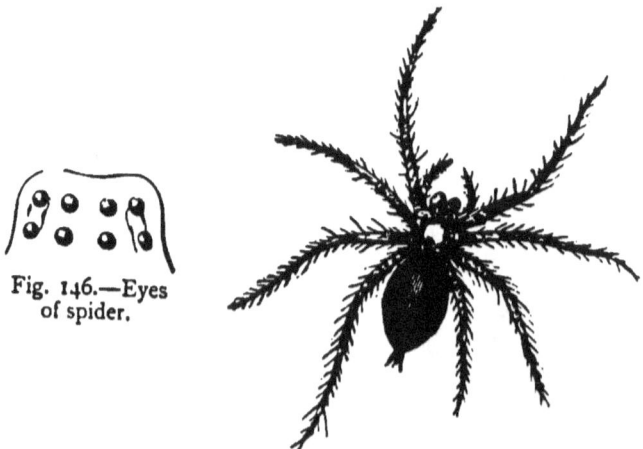

Fig. 146.—Eyes of spider.

Fig. 147.—*Tegenaria atrica*.

ordinary conditions, look at the water-spider (*Argyroneta aquatica*), a spider which breathes air

Fig. 148.—Garden spider (*Epeira diadema*).

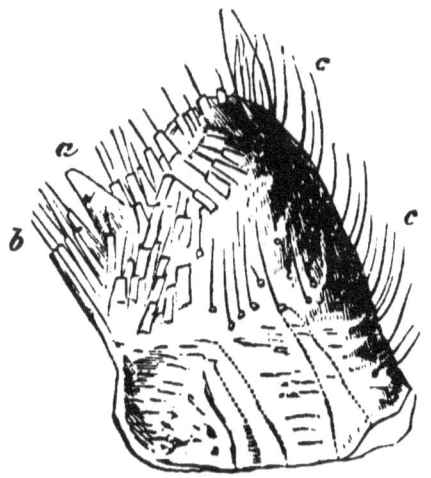

Fig. 149.—Spinneret of garden spider (magnified 100 times).
a, Tubes; *b*, hairs; *c*, sac.

and yet lives at the bottom of ponds and rivers, where it makes its closely woven silken tent like a

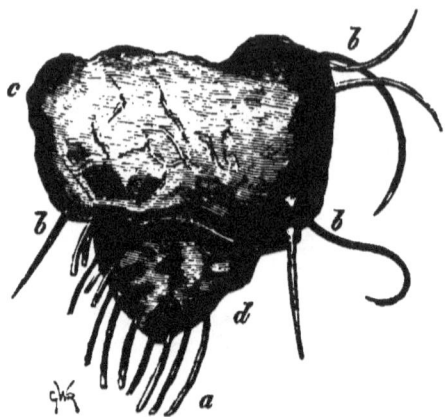

Fig. 150.—Spinneret of gossamer spider (magnified 100 times).
a, Tubes; *b*, hairs; *c*, sac.

Fig. 151.—Nest of spider (*Theridion riparium*)

diving-bell, and fills it with air from above. You can see its body in the aquarium, looking as if

Fig. 152.—Water-spider (*Argyroneta aquatica*), male.

coated with quicksilver. That is the film of air it purposely entangled up above, and has thus conveniently brought down to fill its aquatic tent with.

Fig. 153.—Water-spider, female.

The boys soon learned what a capital hunting-ground lay beneath the looseish bark of any old

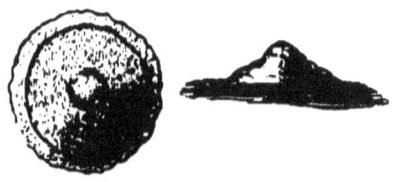

Fig. 154.—Cocoon of water-spider.

tree. One day, Jack's cousin found a prize—a singular-looking creature, looking like an animated microscopic - scaled feather. Examined with a two-inch power, the exquisite beauty of the object produced vocal outbursts of admiration. The surface and margin of its body were adorned

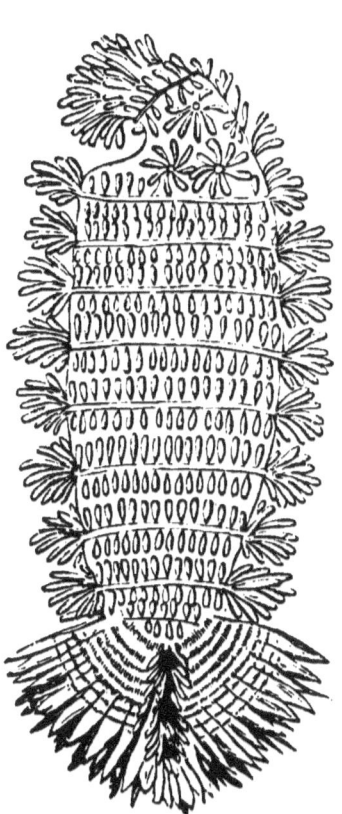

Fig. 156.—Hairs or feathers on tufts of Polyxenes (magnified 250).

Fig. 155.—Hare-tailed millipede (*Polyxenes lagurus*), magnified.

Fig. 157.—Hairs of tail of Polyxenes (magnified 200).

with scales and tufts of feathery hairs. When these hairs were more highly magnified, the accompanying illustrations will give my readers some idea of what the lads saw.

It ran about on its twelve pairs of tiny legs as if they were castors, or wheel skates. This peculiar animal is nearly related to the Julus before pictured, and the centipedes and millipedes generally.

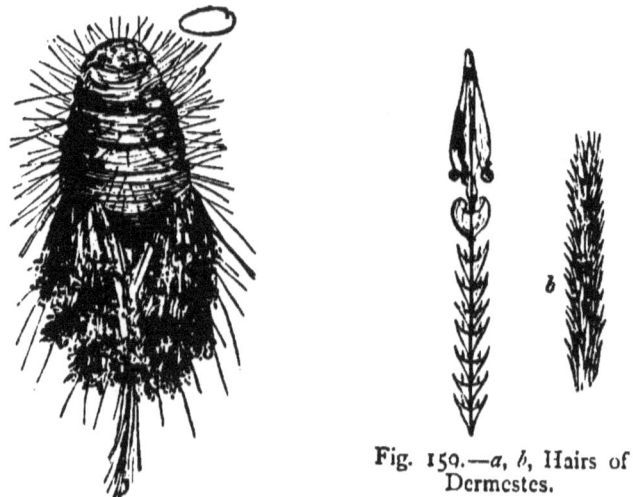

Fig. 159.—*a*, *b*, Hairs of Dermestes.

Fig. 158.—Larva of beetle (Dermestes) covered with compound hairs.

Singularly enough, the same habitat, viz. the protection of a tree's bark, is the dwelling-place of the grub or larva of a peculiar beetle, Dermestes, which assumes a similar appearance to that of feather-tail (*Polyxenes lagurus*). It also is covered with fine hairs, some of which are feathery.

There was an awful rush one sunny, dewy morning, lads and lasses competing, as if there were no such thing as sex, who should be first to carry certain news and a certain object to the scientific Jack. It was a great green locust-like object, clinging to the greenest part of a green shrub, and

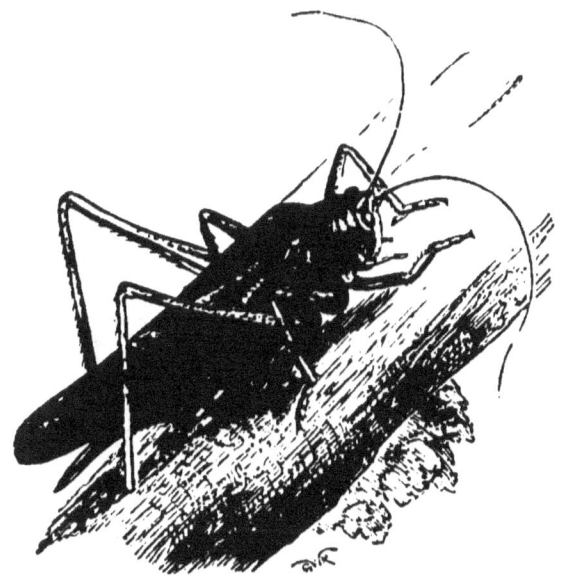

Fig. 160.—Great green grasshopper (*Acridia viridissima*).

so resembling it in the tint and tone of its greenness, that none could doubt its colour was as protective as if it had resembled a wasp or hornet, and had been mistaken for those much-avoided insects. It is a common insect enough, especially in the latter part of summer; but few people notice it on account of its vivid green screening it from observation. I have known this creature to be kept as a pet. It is the great green grasshopper (*Acridia viridissima*).

One day, a cousin happened to lift up a brick in the garden (which was a very old one), and found it swarming with bluish-black "lice," as he

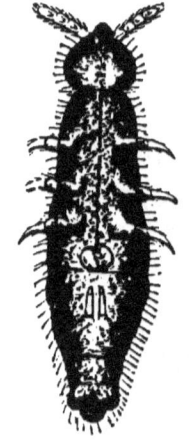

Fig. 161.—Under side of *Achorutes purpurescens* (magnified 32), one of the spring-tails.

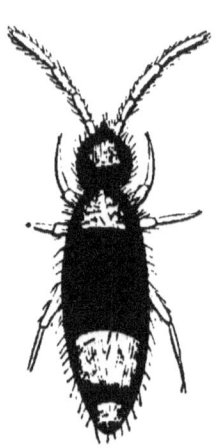

Fig. 162.—*Degeeria cincta* (magnified 25).

called them. People call everything lice that swarms in abundance, and about which they don't know anything. The other day a man brought me a bottle of coloured water from his pond, which "swarmed with lice," he said. It swarmed with Volvox, a remarkable and almost microscopical form of water-plant, one of the most interesting and harmless in the world. But the man evidently thought these " lice " would poison his sheep if they drank of the water.

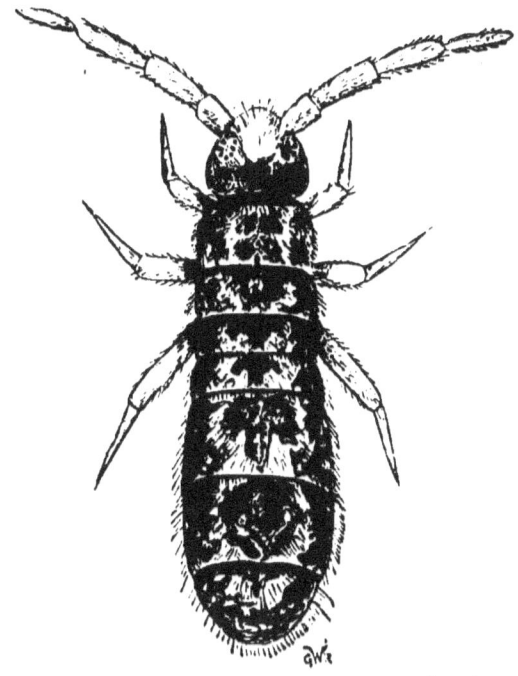

Fig. 163.—Podura (without scales), common under stones (mag.).

It is high time men were not such fools as their ignorance causes them to be. Like all others of that genus, they are mercifully protected from knowing they belong to the "majority."

It so happened that the cook, who knew and loved the children well, notwithstanding what they considered "her haughty ways" (when she would not allow them to eat what they liked)—it so happened, I say, that she found a lot of silvery-looking "hinsecs" sliding and gliding about the cold lamb and the loaves in her pantry. They were fish-moths.

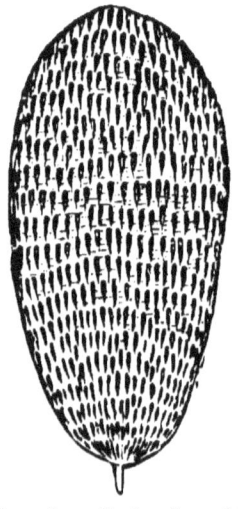 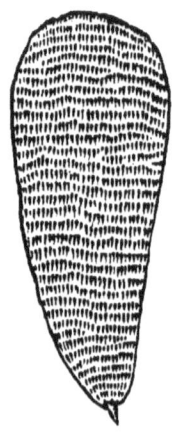

Fig. 164.—Scale of speckled Podura (highly magnified).   Fig. 165.—Scale of black Podura (magnified).

"Look here, Marster Jack," she said, "here's some varmin for ye, an' I wish you'd kill the lot for your microscope."

There was a tin pepper-box full of them. I suspect they had been purposely imprisoned in that abode of spicery out of revenge. Perhaps the cook thought the insects would sneeze themselves to death.

These insects belong to the remarkable order known as Thysanuridæ. Most of this order have bodies covered with scales, not unlike those on butterflies' wings, and these scales are so prettily

Fig. 166.—The speckled Podura (magnified 30).

marked that they have long been used as tests for the accurate definition of good microscopes. They are found in damp places generally—damp cellars, damp walls, damp pantries, on the surfaces of weedy ponds, in greenhouses, under stones, in empty flower-pots, beneath the bark of trees, etc. Many of them go by the popular name of spring-tails, for when disturbed they jump like fleas. Several kinds are peculiar to the seaside, such as Lipura. The black Podura are especially abundant in cellars, and if you want to catch them you

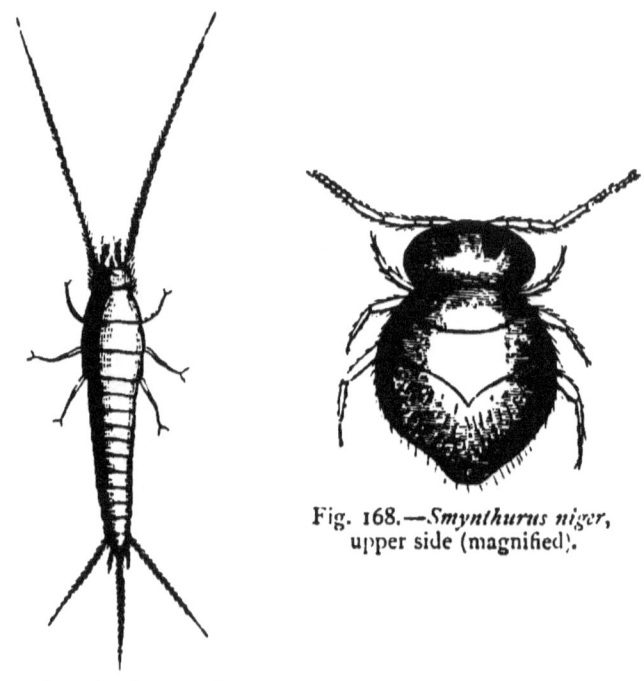

Fig. 168.—*Smynthurus niger*, upper side (magnified).

Fig. 167.—*Lepisma saccharina* (magnified).

have only to set a trap in the shape of a cold stale potato or an old mutton-bone. These Podura have sixteen eyes, set in two groups, one on each side the head, looking to all the world like a cluster of beads. Their tails double up under the body something like the wooden toy frogs, which have a spring and bit of wood fastened by cobbler's wax, so as to give way and jerk the toy up when the wood becomes unfastened. They are all wingless insects. It is capital fun to find their eggs and watch them hatch out.

CHAPTER VI.

LAND SHELLS.

THREE weeks after the holidays commenced, Jack received the following letter from his friend Willie :—

"Bromlea, August 16, 1888.

"DEAR JACK,

"How are you getting on? I miss you awfully. What jolly times we all had the last term! I was down in the dumps when I got home, and didn't know what to do. My dear old dad saw what was up, and persuaded me to go out for drives with him when he visited his country patients. He is the dearest, gentlest, old dad in the whole world, and he soon got out of me what you and I had been up to—all about our rambles and collections, and so on. It quite did me good to tell him; he seemed so interested, and it was very pleasant to go over the old ground again.

"As you know dad is awfully fond of natural

history. I always knew that, but I never found out, before these drives of ours, what a real blessing it had been to him during his lonely country drives —how he observed the birds and their songs, the insects, the flowers, and a thousand objects besides. As we drove along, he was full of these things, and the green country lanes seemed like an open book to him—a book he never got tired of reading.

"He has been showing me how to work the microscope, and how to mount specimens for examination. It isn't half so difficult as I used to think. Dad was quite pleased when he saw I took a real interest in the business. I always liked to potter about these things, as you know; but now, somehow, I feel more serious about it.

"'I should like to go in for something special these holidays,' I said to him. 'What should I take up? Something special to collect and arrange, and get to know about—something other than butterflies and moths, you know,' said I.

"'Very well,' he replied; 'begin collecting snail-shells.'

"'Snail-shells!' said I, somewhat disgusted—for I hate snails and slugs.

"'Why not? You were very fond of collecting

sea-shells when we went to the seaside last year; why not collect land-shells, and fresh-water shells too, if you get a chance? You have no idea how interesting it is.'

"'But there are not many sorts to collect,' said I. 'It won't take me more than a day or two to get all the kinds there is about here.'

"'Oh, won't it?' said he. 'Just you try. You will be very surprised to find what a number there are, and how they want looking for.'

"Well, you know, that kind of started me. Still, somehow snail-shells seemed rather a mean kind of things to collect. They are such slow-coaches, and therefore I fancied I should have to hunt them slowly.

"I thought it over a day or two, and old dad never let the subject drop, but referred to it every now and then as we drove along. Then I got to know from him what interesting things both land and fresh-water snails are—how ancient are their shapes and habits; how they have been living in the world for millions of years, sometimes they had actually formed *marble* by their accumulated shells, as the well-known Sussex marble and Purbeck marble; and how, in the hollow fossilized tree-trunks of gigantic club-mosses called Sigillaria, which grew when coal was forming, there had been found land-shells almost exactly like the little delicate Pupas we find in the moss of the hedgebank. I had no idea before that snails had such an ancient history.

"Then the dad told me how I should find different species in different habitats. He explained that the last word was much used by naturalists, as expressing the natural conditions which surrounded any living object. He said I should find some species of land-snails living under one set of conditions, and another species under quite a different set. Perhaps, he hinted, it had been these different circumstances surrounding them which had helped to give different species their leading characters. He told me I should find some which liked limey soils and rocks, and others which did not; some which loved to live in damp places, and others in dry; some of them nearly as big as my fist, and others not much bigger than a pin's head.

"Well, I won't bore you any further; but the upshot is, I am now collecting snail-shells, and jolly fun it is too, I can tell you. Dad laughs a quiet laugh sometimes, now that he sees I am so dead on snails. I don't mind, for if it hadn't been for him, I should have had the same duffing notions about them now I had when he set me to work. Good-bye. Let's know what you are up to.

"WILLIE."

A good many letters passed between the lads, notwithstanding schoolboys are not good correspondents—except when they want something.

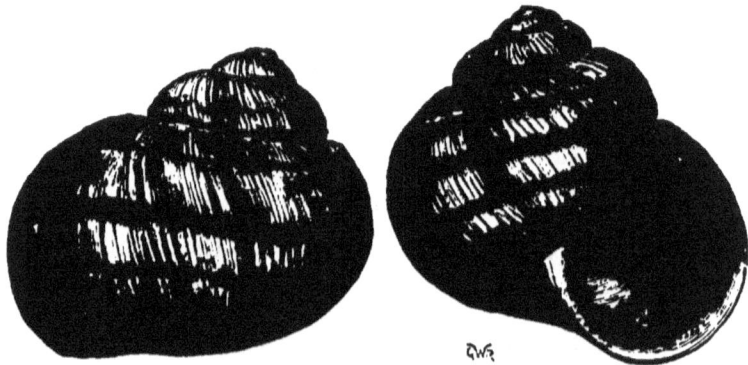

Fig. 169.—*Helix pomatia.*   Fig. 170.—Ditto, mouth view.

But I need not repeat them, nor give them in full Suffice it that Willie became a keen hunter of land-shells before the holidays were over, and managed to make a capital collection, including some rare ones. The long journeys his father was obliged to make enabled Willie, whilst waiting for him, to hunt in new and varied localities. Of course, the big kinds, such as *Helix aspersa*, and the common garden and wood snails (*Helix hor-*

tensis and *H. nemoralis)* were soon sought and found. The young hunter was perfectly surprised at the exquisite colours and variation of colour-tints the two latter species possessed.

His father showed him how and where to look for the rarer and smaller kinds—pulling up plants by the roots and shaking them; turning over the damp, rotting leaves in the wood; grubbing beneath

Fig. 171.—*Helix aspersa.*

Fig. 172.—*Helix ericetorum.*

clumps of ferns and wood-rushes for the smaller Helices, pupæ, etc.; closely examining the trunks

Fig. 173.—*Helix nemoralis.*

of trees for Clausilia, Bulimus, and special kinds of Helix. It was in this way he found *Bulimus obscurus*, whose shell was almost entirely *obscurea* by the dirt the cautious creature had covered itself

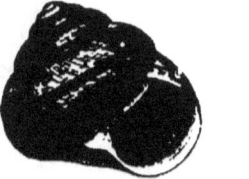

Fig. 174.—*Helix arbustorum*.

with, to escape detection. Beneath the rotting bark of trees, he taught him to look for and find Balea ;

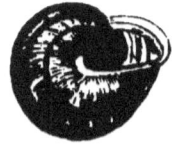 underneath stones, Helicella ; along the sanddrives of the coast (and inland where the soil was limey), *Helix caperata, Bulimus acutus*, etc.

Fig. 175.—*Helix hortensis*.

Willie also learned that the best time for collecting land-shells is the autumn, when they are fully grown, and are most beautifully marked. Those

Fig. 176.—*Helix virgata*.

collected in spring are usually winter-worn and weathered specimens.

This newly found fad of land-shell collecting

was a capital excuse for Willie (who was an affectionate lad, and couldn't bear to be long without

Fig. 177.—*Helix cantiana.*

hearing or seeing anybody he liked) to write to his "professor," asking him how he was to

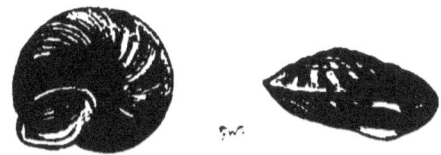

Fig. 178.—*Helix lapicidia.*

prepare the specimens he had captured for the cabinet.

The professor was in Switzerland at the time,

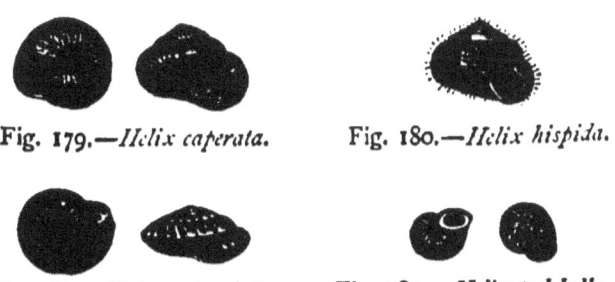

Fig. 179.—*Helix caperata.*   Fig. 180.—*Helix hispida.*

Fig. 181.—*Helix rotundata.*   Fig. 182.—*Helix pulchella.*

and Willie's letter reached him there. Notwithstanding the bother of writing letters when one is away from home, and supposed to be enjoying one's self, the professor answered Willie's letter by

Fig. 183.—*Helix rufescens.*   Fig. 184.—*Helix lamellata.*

the same day's post. Truth to tell, he regarded Willie and Jack as intellectual children of his own.

Fig. 185.—*Helix pigmæa.*   Fig. 186.—*Helix Carthusiana.*

He had been the chief means of opening their minds to the abounding works of the Great Father; therefore the lads were in a sense his own intellectual progeny. It is this intellectual affiliation which knits the true master and the true student so closely together.

Fig. 187.—*Helix aculeata.*

So, partly for the sake of writing to Willie, and chiefly to help him, he wrote to him as follows :—

"When you collect your living specimens, keep them a few days. Then remove the animals from the houses (shells) they have made. At first this appears a nasty sort of a job. But all real practical natural history, dissecting, etc., seems nasty at first.

I dare say the first man who ate an oyster was thought a 'nasty man' by those who looked on, being too timid to attempt the task themselves. I am very sure the English people would think those dwelling in the south of France 'nasty,' if they were to see the great heaps of the Roman snail (*Helix pomatia*)—a British species, by the way, although probably introduced by the Romans, who were fond of snails—and our commoner English large snail (*Helix aspersa*) offered for sale in the fish-markets, and eagerly bought. If they tasted the *pâtés* made of them, however—that is, if they didn't know what they were eating—they would give up the silly charge of 'nastiness' about anything a body did not happen to like.

"Well, the first thing you've got to do is to separate the living snail from its house. Remember that a snail-shell is as much a part of a living mollusc as a lobster's crust, or as your own bones are part of your own living self.

"When you kill the snail in its shell, you sever the connection between the two. The step is to remove the animals, and this may not seem a pleasant job. Kill them by plunging in boiling water, when the muscular connection with their shells will be severed. Then the body can be easily taken out with a pin, exactly like picking periwinkles.

"With regard to the smaller kinds, whose shells may not have mouths wide enough to admit even a pin for picking purposes—indeed, such shells as Clausilia, Bulimus, Helicella, etc., are too little to allow of much handling with a pin, even if the animal had not drawn itself up into the extreme corner of the shell the moment it scented danger —the best thing to do with them is to let their little bodies dry up in their shells.

"When you have picked out the animals from the larger shells, wash out well with warm water, and then place them before a fire to dry. Don't rub them, but clean off any dirt that may have remained with a dry camel's-hair pencil. Some of the small snails have their shells covered with hairs, or short bristles, and you must mind not to remove them."

Willie's father delighted to induct his lad into the fragments of natural-science lore he himself once possessed—"broken lights" of other days, when both money and leisure were more abundant, and before the household became so thronged with curly-haired tyrants.

Alas! the theory of evolution, microscopical

dissection, embryology, phylogeny, and a host of other deeply interesting and still more deeply complicated subjects had grown up almost unconsciously while the village doctor had been toiling for the crickets on his hearth. Nevertheless, not a little (and that little was good) remained of the earnest days of his younger manhood.

A man possessed of a son gets a double chance out of life, particularly if the boy inherits his father's tastes. Then he lives again, renews his youth, enters the glorious lists of young manhood a second time—rejoices when his son succeeds, mourns when his son fails.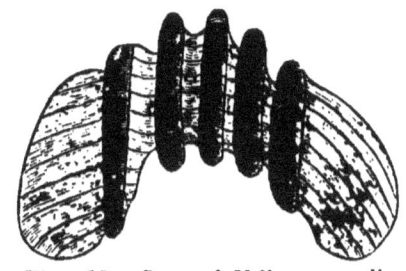

Fig. 188.—Jaws of *Helix nemoralis* (magnified).

You cannot do that with your daughters, although they are " ever so much nicer!"

So when Willie was pulling the snails out of their shells with a pin, his father showed him there were several other things he might do at the same time. Thus, every snail has a more or less special kind of *jaw*, used for feeding purposes. Then they also possess *odontophores*, or tooth-bearing straps, all thickly set over with silicious teeth, which gleam and glisten like precious stones when examined by polarized, or partly decomposed, light. The true teeth of snails, and indeed of all univalve mollusca, are therefore not in the mouth. But the jaws of snails are, and their duty is to assist the odontophore by triturating the food. These jaws, and the number of the ridges on them, are now being carefully studied. They are not limey, but

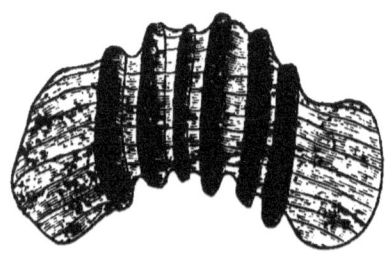

Fig. 189.—A second form of ditto.

chitinous — that is, formed of a substance like horn, or one's own finger-nail. Willie and his father, therefore, added a new delight to snail-shell collecting. They referred to a chapter on the subject in a natural history magazine, which the latter had carefully taken in from its beginning, and to whose blue-cloth bound volumes he always turned when in a difficulty—I mean *Hardwicke's Science-Gossip*.

The reason was that Willie wanted to send to Jack full details of how to proceed in dissecting land and fresh-water snails for the sake of their jaws and odontophores. So he copied the following paragraph from a page his father had turned down for him :—

"For the method of dissecting odontophores, jaws, etc., of molluscs (which should be done under water, in a white shallow dish), almost any book on the microscope will give you full information. I advise the simple method of dissecting the animal with forceps and needle, to the common one resorted to by many conchologists of boiling the animal if small, or parts if large, in sodic or potassic hydrates, on purpose to procure the jaw or lingual ribbons. The attachments are often appended, and the object has a more natural appearance under the microscope. The jaws of the small *Helix virgata* can be seen with the naked eye in the dissecting-trough, and the smallest species may be crushed and washed in the sunken cell of a micro-slip, using a two-inch objective for detection.

Fig. 190.—A third form of ditto.

"As the jaws are found, place the different kinds in watch-glasses or small colour-saucers, until they are so dry that they can be transferred to small pill-boxes without any risk of their sticking to the bottoms or sides, and so of carrying foreign matter. There they are dust-free, and can be stored for any length of time. On the lid of each pill-box should be written such information as would be required for a micro-label slide. This method of preparation is not applicable to odontophores. In a day or two the jaws will have become thoroughly dry. After soaking in turpentine, they may be mounted on micro-slips in any soluble form of Canada balsam, and without a cell. A wire-clip will hold in position for a few days, until there is a little 'set' in the medium. As evaporation takes place, fill up with fresh balsam. When dry, 'ring' the slides twice with a thick solution of dammar in benzole, and varnish. Any number of jaws of molluscs, agreeing in character or shell-colours, may easily be mounted on one slip; three or more are easily treated, and, with a little care in balancing the clip directly over the specimens, it will be found practicable to mount one or two jaws without a cell. Canada balsam is much better for this work than glycerine or similar fluids. It is more easily manipulated, requires no extra care in fastening up, and is handy for polariscope work."

Willie's father pointed out the beginnings of the doctrine of "natural selection," as proved even by snails' jaws. Thus the ridges or cross-bars of *Helix nemoralis* vary from two to four in one locality. In another (on the limestone), they range in number up to seven ridges or cross-bars. It is the same with the garden snail and other common

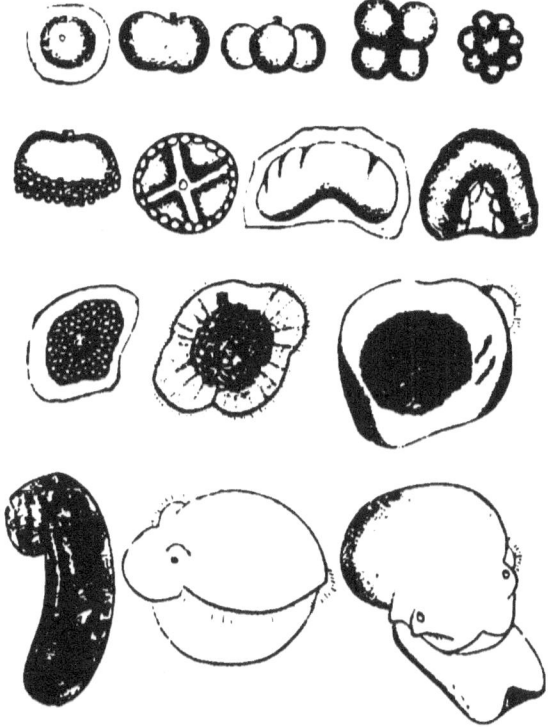

Fig. 191.—Successive stages in the development of a fresh-water snail (*Limnæa pereger*).

species—the ridges in the jaws vary in number with the kind of vegetable diet they affect; and this, of course, is determined chiefly by the characters of the subsoils and geology generally.

August happened to be rather a dull month that year. It was cloudy when it was not rainy, and the days seemed to draw in on purpose to make the evenings longer. So the microscope became an evening toy, as well as a scientific instrument. Then Willie learned how the eggs of snails, especially the fresh-water kinds, such as Limnæa and Paludina, hatched out larvæ at first resembling infusorial animalcules in their possession of *cilia*, or eyelash-like hairs arranged around the mouth ; how they regularly rotated within their cells until they were set free to move through the water by the same means ; and how very possibly (only it was for the young and rising generation of naturalists to determine the fact) the embryos of landsnails did the same thing within their eggshells, only more expeditiously, being pinched for time, and therefore obliged to make overtime within their eggs—to *accelerate* their larval stages, in short.

CHAPTER VII.

"THEY GO A-FISHING."

"I GO a-fishing," said St. Peter, when one of the other disciples wanted to know what he was about to do for the next twenty-four hours. Unfortunately, it is your *idle* friends, as a rule, who want to know how you mean to spend the day. They have not the slightest idea how they are going to spend it themselves; so, at any rate, there is some chance of their getting a little interest out of life by knowing beforehand how you mean to spend yours. If they cannot join you in body or spirit, at any rate the derived knowledge leaves them at liberty to criticize you. And that is all that very weak people can do.

Imagine Jack's delight (which was contagious, especially among the girls, to whom Jack had laid bare his heart's idol) when one evening, just before supper, a limping lad, dressed in the garb of the E. T., rang the front-door bell, and looked as weary as he could, for he knew that would secure a supper. He had been to the hall with telegrams before!

It was a telegram for Master Jack. We apply that term to boys now, but it was a grand name given to valiant fighting-men in the days of old Border warfare.

Master Jack didn't mind the mild chaff of those around him, who would have given their only half-crown to have got a telegram all to themselves. He ripped it open with a knife deliberately, as if he had been in the habit of receiving telegrams until he was bored with them.

Then he passed it quietly to his sister, who flushed as she read it, announcing that Willie would be with the lively party by noon the day following.

Though Jack passed the telegram quietly, he was pleased beyond measure. Between him and Willie there had been established the strongest brotherhood in the world, infinitely stronger than the brotherhood of bodies—the fraternity of *souls*.

However, I am not a novelist, or I should devote a chapter to their meeting, and describe how Willie was flurried on being introduced to Jack's sister, and how Jack's sister blushed as red as a peony, and hated Willie all the more because she knew she had done so, and couldn't help it.

So I leave the lads alone, to compare delightful notes, and still more delightful experiences. All I have to do is to be the humble chronicler of their next and newest set of natural-history adventures.

They determined not to collect butterflies, moths, bees, wasps, dipterous flies (unless, of course, they saw some "jolly rare ones"), beetles, snails, slugs, etc. Willie observed, "I say, let's go dredging!"

Now, that was a capital thought; none of them had ever thought of *dredging*. I doubt whether most of them knew what it meant. But they applauded the idea—for that is the best way of not being thought ignorant of it.

It was ultimately deemed best, however, to hunt for water-insects, and odds and ends of aquatic life. It is remarkable what a number of creatures belonging to widely different orders agree in possessing a common habitat. Depend upon it, either a great many changes have taken place in water-insects since they first appeared on the earth, or else in the larger numbers which swarm the atmosphere. There is reason to believe that, ages ago, some of the earliest forms of insects were aquatic; and it may be that the living species inhabiting fresh waters all over the world are still maintaining their ancestral habits. Some species, however, such as the water-beetles, must certainly have acquired the habit of being able to live in water, for their practice is so different to the rest of their tribe.

The surface of the ponds and tarns visited by Jack, Willie, and the other young friends, where

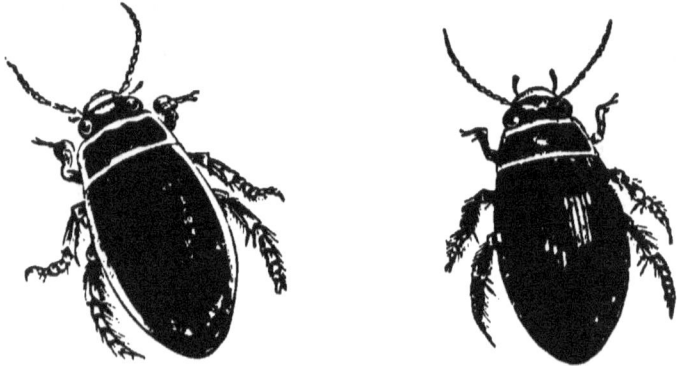

Fig. 192.—Water-beetles (*Dytiscus marginalis*), male and female.

not covered over by pond-weed, duck-weed, and frog-bit, looked smooth and dark. The open spaces were so many skating-rinks to the whirligig beetles, which glided gracefully about; and although it looked as if some of them must come into collision, they never did. They beat skaters in that respect. It seems a strange whim for these glassy little water-beetles to do nothing all day but waltz about in this ridiculous fashion. What do they do it for? But perhaps they would ask a similar question of ourselves, if they saw a hundred men and women taking a delight in skating all day.

Then there were the huge beetles *Dytiscus marginalis* and *Hydrophilus piceus;* the former, both in its larval and adult stages, one of the most voracious of creatures, as all who have kept fresh-water

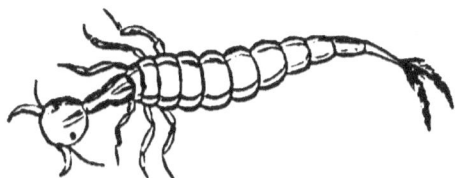

Fig. 193.—Larva of *Dytiscus marginalis.*

aquaria know. The full-grown beetle will attack small fish, frogs, newts, with impunity. Nor is the larva much less to be dreaded for its voracious appetite. They are to tadpoles what the wolf is to a flock of young lambs.

The peculiar way in which the Dytiscus manages to breathe the air dissolved in water is best seen by examining the air-apertures, or spiracles, which are both beautiful and instructive objects when mounted. The *tracheal* or breathing system is very complicated. You get a dead beetle and make a careful incision down the back with

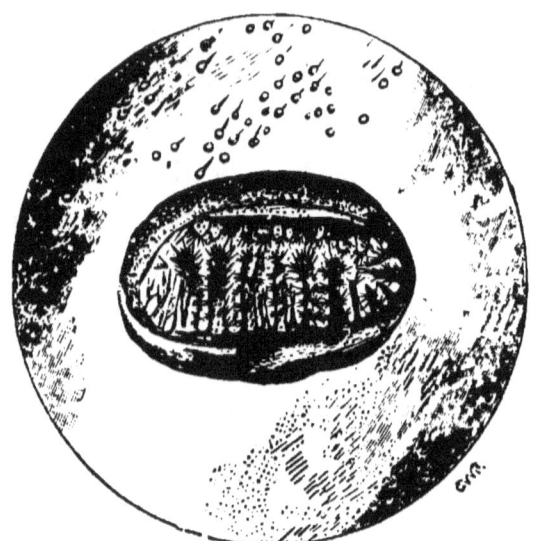

Fig. 194.—Terminal spiracle of Dytiscus.

Fig. 195.—Parasite of Dytiscus.

scissors; then soak the entire object well in acetic acid, and use the forceps to detach the trachea,

which lie on either side the body, and communicate with the air by means of the gilled spiracles.

But, greedy as the Dytiscus is, it is the prey of a parasite, which may usually be found deeply imbedded in the skin near the spiracles, as you are dissecting.

Fig. 196.—Cocoon of Hydrophilus.

The Hydrophilus is almost as voracious as its neighbour and competitor. The boys found a female in the act of depositing its eggs in the nest. They noticed how the nest was closed up until it resembled a glass retort. There the eggs were safe and sound, and there they hatched out into "greedy larva not unlike those of Dytiscus.

Another group of common aquatic insects allied

to the water-scorpions (Nepa) were discovered—*Ranatra linearis*—and transferred to the wide-mouthed pickle-jars for further observation. It was amusing to see them take their prey, assuming a position not unlike that of the praying mantis. Even more interesting was the discovery of the eggs of this singular creature, found arranged in

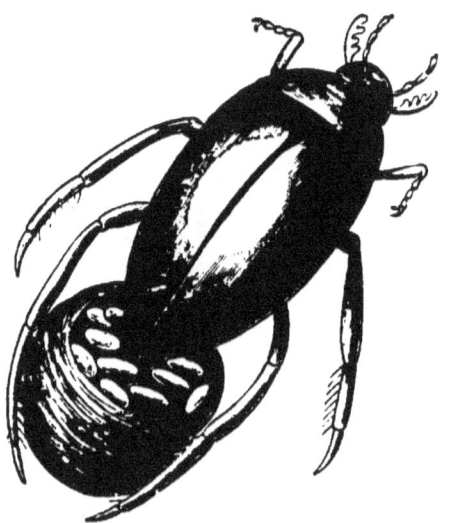

Fig. 197.—Another water-beetle (*Hydrophilus piceus*), in the act of depositing its eggs.

a row, penetrating the leaf of a frog-bit (Hydrocharis), each egg having a couple of antennæ-like processes to prevent its slipping through. Out of these funny eggs young Ranatræ were hatched, and wonderful skeleton-like things they looked.

"THEY GO A-FISHING." 149

The bottoms of the ponds and streams visited during the long sunny mornings were seen occupied by slowly moving objects, which, how-

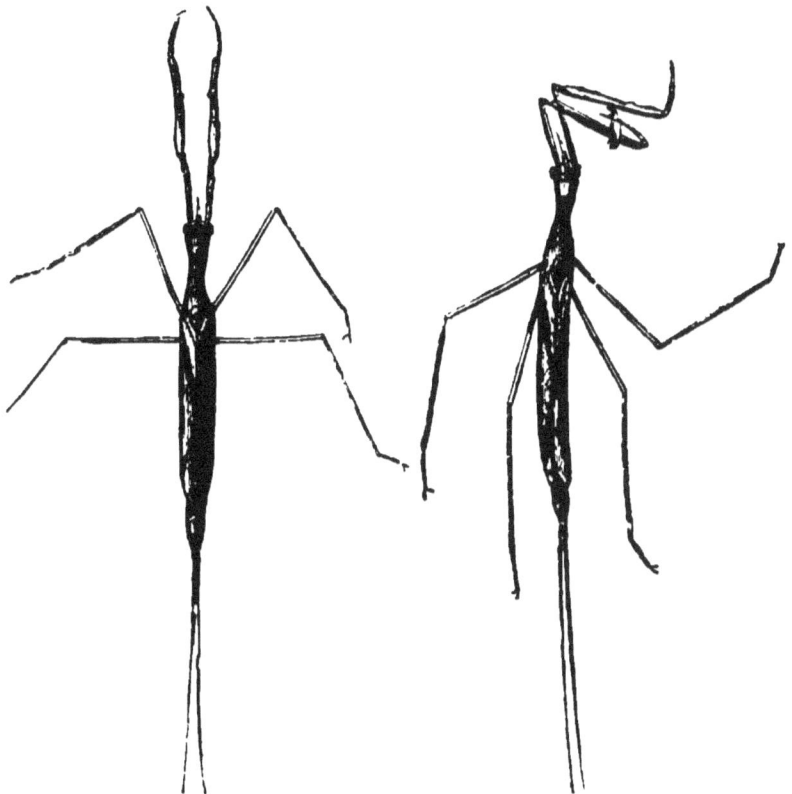

Fig. 198.—Full-grown *Ranatra linearis* (magnified). Fig. 199.—*Ranatra linearis* in the act of catching its prey.

ever, were very erratic in their movements. The greater number of them looked like animated sticks; some were clusters of minute Planorbids and other fresh-water shells. An examination showed them to be different species of caddis-worms, the aquatic larva of the Ephemeridæ, or

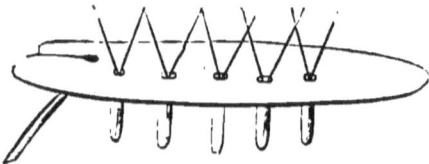

Fig. 200.—Eggs of Ranatra deposited on leaf of frog-bit (Hydrocharis).

may-flies. It appears very singular that numbers of air-breathing and winged insects should pass

Fig. 201.—Single eggs and young of Ranatra.

Fig. 202.—Larva of Ephemera, or caddis-worm fly.

their earliest stages in the water. The breathing-organs of the larva of Ephemera are very beautiful when seen under the microscope; and the ingenious way in which these aquatic breathers are converted into an air-breathing apparatus when the creature leaves the water for good, is more or

Fig. 203.—Case of caddis-worm (*Phryganea grandis*).   Fig. 204.—Shell-case of *Limnephilus flavicornis*.

less the same as marks the transformation of the dragon-flies, gnats, etc. The transformation of a gnat's egg into a free-swimming larva, and afterwards into a winged fly, is well known to naturalists. You cannot wonder that may-flies, gnats,

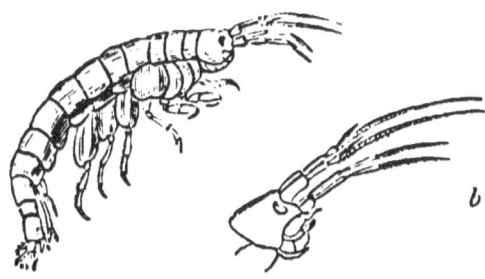

Fig. 205.—Fresh-water shrimp (*Gammarus pulex*). *b*, Antennæ of ditto (magnified).

and dragon-flies haunt the streams, ponds, and dykes, even when they have become winged. It is in the water rather than in the atmosphere that the greater part of their lives is passed. One species of insect actually never leaves the water, even when it has acquired wings, but uses them as swimming-organs.

Then, again, in most ponds and natural streams you find creatures belonging to a widely separate division — the Crustacea. Moreover, some

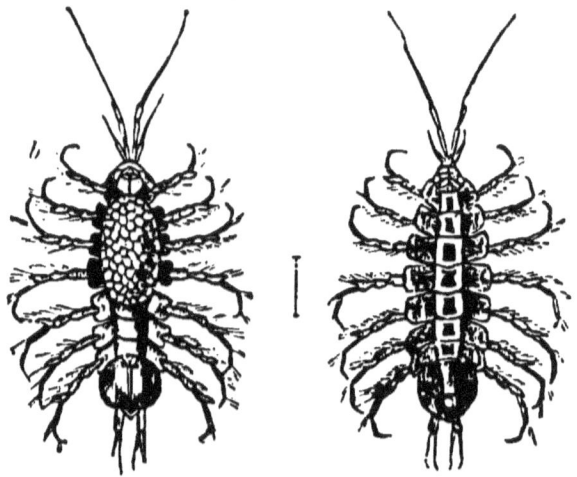

Fig. 206.—*Asellus aquaticus.*

of these fresh-water forms are nearly allied to those now living in the sea, such as the freshwater shrimp. Others are not distantly related to the Trilobites which swarmed in the seas of the Silurian period, many millions of years ago, such as *Asellus aquaticus*, found in most boggy tarns.

Then there are *worms*, some of which may also be derived from ancestors that formerly lived in the sea. Of course the lads looked out for and caught specimens of the curious hair-worm (*Gordius aquaticus*); what boys would not? For is it not an article of belief among schoolboys that these hair-worms are produced by taking a long hair from a horse's tail, and placing it in a saucer to stand in the sun? Boys believe so, and their grown-up ancestors also believed it, without thinking they were committing themselves to the atheistic

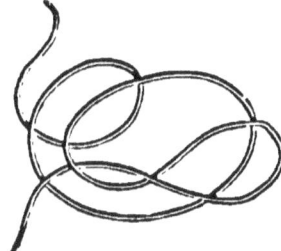

Fig. 207.—Hair-worm.

Fig. 208.—End of hair-worm.

doctrine of spontaneous generation. Naturalists, however, know that this worm undergoes its first changes in the bodies of insects, chiefly beetles, and that it has nothing to do with the horse or any other animal's hairs, except its marvellous resemblance to them. If you were to behold one of these hair-worms, when imprisoned in the zoophyte-trough, discharging its eggs, you would see a wonderful sight. They are poured forth in thousands, until one is bewildered by a new sense of abundance.

Many of the worms which permanently live

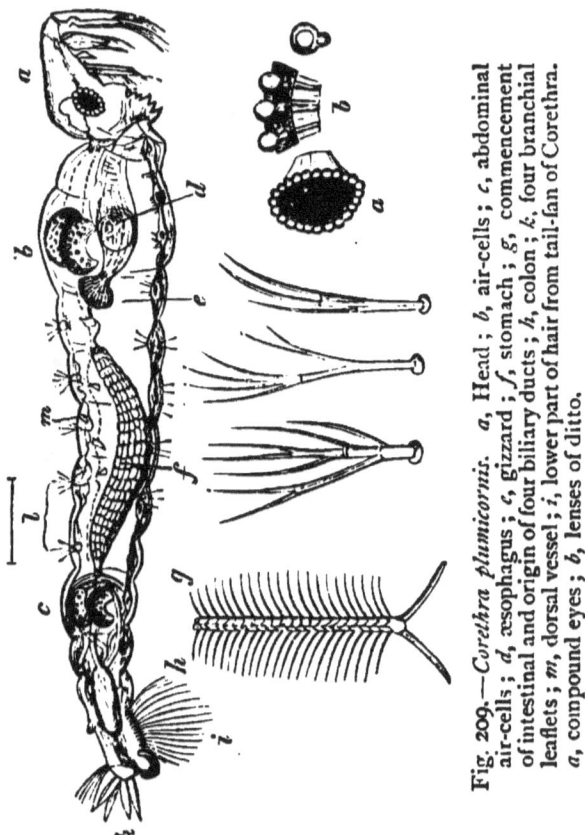

Fig. 209.—*Corethra plumicornis*. *a*, Head; *b*, air-cells; *c*, abdominal air-cells; *d*, æsophagus; *e*, gizzard; *f*, stomach; *g*, commencement of intestinal and origin of four biliary ducts; *h*, colon; *k*, four branchial leaflets; *m*, dorsal vessel; *i*, lower part of hair from tail-fan of Corethra. *a*, compound eyes; *b*, lenses of ditto.

in fresh water possess the usual transparency characteristic of nearly all the lowly organized aquatic animals, so that you can plainly see the internal organs. One of the most beautiful and instructive in this respect is the not uncommon *Corethra plumiformis*. Some of these worms so nearly resemble the larvæ of aquatic insects, or rather, the latter appear so often like the former that the young naturalist is frequently bewildered in distinguishing one from the other.

Corethra is a splendid study, on account of its *hyalinity*—its glass-like tissues. You can see its compound eyes, air-cells, gizzard, stomach, intestinal arrangement, and *branchial* or breathing arrangements, almost as easily as if you were looking through a window at the contents of a shop —where, of course, all the best things are arranged.

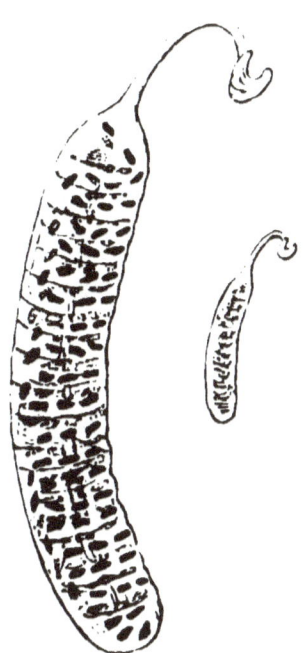

Fig. 210.—Egg-bag of common gnat (*Tipula plumicornis*), natural size and magnified.

One of the most interesting set of captures and observations made were on the *common gnat*, which, as everybody knows, deposits its eggs in water You can get any quantity of them in any exposed rain-water tub.

The female gnat lays her eggs early in August, as a sausage-shaped bag of nearly colourless jelly,

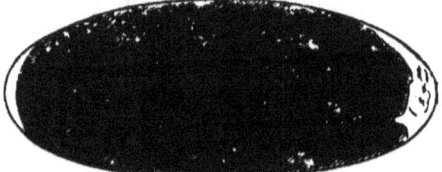

Fig. 211.—Egg, first day.

Fig. 212.—Egg, five hours later.

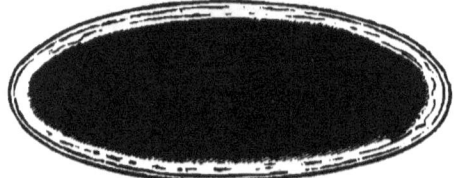

Fig. 213.—Egg, second day.

Fig. 214.—Egg, five days old.

varying from a quarter of an inch to one inch in length, beneath the surface of still water. The egg presents in profile a convex upper and a flattened lower surface (these terms, upper and lower, are applied simply for convenience). Looking down upon it, the appearance is oval. The shell is

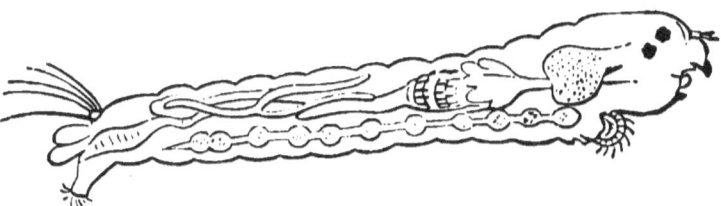

Fig. 215.—Diagrammatic figure of young larva.

somewhat elastic, and is very transparent and structureless, so that the development of the embryo may be easily observed. The mass of

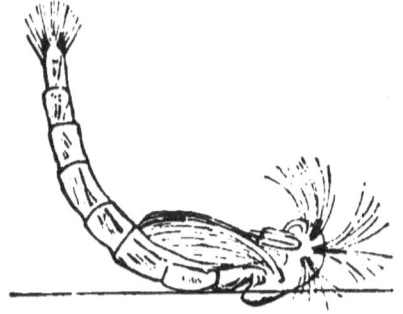

Fig. 216.—The nymph.

eggs have a sucking-disk at one end to fix them by. These eggs can be followed through every stage of their hatching; and, thanks to the great transpareney of the egg-shell, or membrane, we can witness the development of the larva within. The first three woodcuts given on p. 156 show the process of alteration within the egg; whilst the last, or fourth egg, demonstrates that the larva is nearly complete. When it is, it bursts through

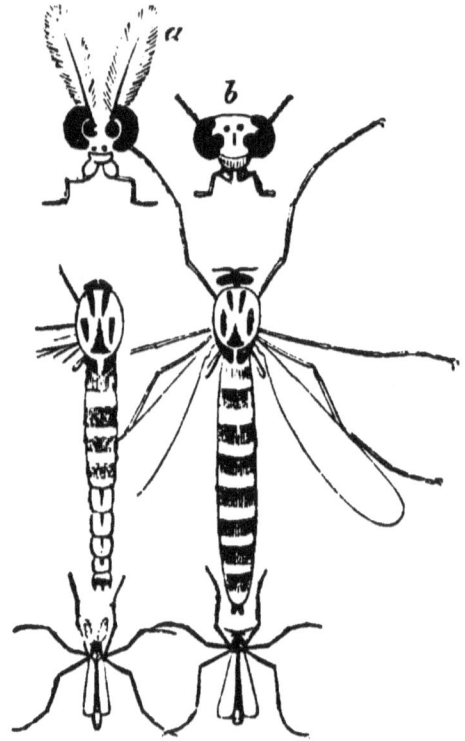

Fig. 217.—*a*, Head of male; *b*, head of female.

the egg, swims actively and curiously about something like a Corethra, and eventually passes into the *nymph* stage—the equivalent to the chrysalis of a butterfly. The hardened skin becomes a kind of "boat cradle," although not of bulrushes, to float the fully developed insect to the surface, whence it may take wing.

But, as a faithful chronicler of young naturalists —than whom there are no people more joyous and optimistic—I cannot stop to mention the great army of commonplace things they collected. "All was fish which came to their net"—that ought to be the motto of every young naturalist. You may be devoted to certain things now—may even regard the collectors and observers of certain other things you don't care for as absurd and even stupid (they perhaps do the same as regards you); but by-and-by, when you grow older, your sympathies will extend, and a fuller and richer life will be the result.

CHAPTER VIII.

A NEW HUNTING-GROUND: AMONG THE MITES.

ONE fine, half-sultry, half-foggy morning, just before breakfast, one of the boys rushed into the room with a piece of bark, on which hosts of creeping things were swarming.

"Look here," said he; "here's a lot of woodlice."

Willie looked at them through his pocket-lens, and saw that all had *eight* legs. Now, legs as a rule can hardly be used as a means of classification, unless for wooden tables. Still, their number is not without value. Thus, if a creature has six legs, you know it is an insect; if eight legs, that it belongs to the spider family (Arachnida); if ten legs, to that represented by shrimps, lobsters, crabs, etc. (Crustacea).

Now, a real woodlouse (Oniscus) is a crustacean, and has ten legs, all of equal length. Willie's glass told him at once the division of the animal kingdom to which the little objects belonged.

"No," said he, "they are not woodlice; they are mites."

"Mites? What! cheese-mites?"

"No," he replied. "Cheese-mites are not the only kind in the world. You have no idea of the number of kinds of mites there are. They can be found almost everywhere, feeding on decaying substances; nearly every kind of plant is haunted by them. They are very common under the bark of trees and shrubs, under stones, and so on. You find them on animals, especially birds, where they devour the waste scurf and feathers, so they are really so many barbers and *plumassiers*, or feather-cleaners. They may be found in nearly every pond, where they are fitted to an aquatic life, and go by the name of water-mites."

That was a long speech for my young friend to make. But it contained news for most of his audience; and, what was more important, it suggested an additional happy hunting-ground.

"I say," eagerly remarked Jack, "let's go mite-hunting to-day."

No sooner said than agreed upon—for boyhood allows no procrastination in the indulgence of its whims.

But, as many of the objects they determined to

hunt are so small that the prick of a pin on paper would about represent their natural size (except the water-mites), they required a good deal of looking for. I doubt whether they would have found some kinds at all, if it had not been for their living together in great numbers (*social* mites), so that they presented the appearance of animated dust. Under the microscope, however, not only could the different species be easily made out, but the males and females of each distinguished—the latter often differing more in appearance from each other than one species does from another.

It is astonishing, after one has commenced a special study of natural objects, how common we find them, although, perhaps, we had hardly suspected their existence before. The scales are removed from our eyes. You soon find out objects when you have learned to take an interest in them. Indeed, the art of "taking an interest" in anything is half the battle.

Thus Jack had heard the gardener talk about " them darned *red* spiders," but he neither asked what they were nor took the trouble even to look at them, until after that eventful morning when the boys went mite-hunting. Then he found that the gardener's "red spiders" were in reality mites.

So away the boys strolled, their pockets crammed with pill-boxes and sandwiches. One of the party got nearly thrown down by a gorse-bush when crossing the common, and this led him to see that the cobwebs entangled in the furze were crowded with a dense red powder.

"Here you are!" he cried. So the powder was pill-boxed, although, somehow or another, it all seemed to run together into a ball. The individual mites were afterwards found to be *Tetranychus telarius*. It seemed that the web was their own, not a spider's. They had somehow spun it for mutual protection or defence against the rain, just as the social caterpillars of the little eggar-moth do, which one sees so abundantly on the hawthorn hedges during a hot and droughty summer.

Many cobwebs are attributed to spiders which, maybe, are the work of these social mites. Now, a spider's web is in reality a trap. Mites' cobwebs are houses, barracks, castles. You will find the lime trees, late in August or early in September, with their trunks and branches often half covered with lovely and delicate webs, on which you see a special kind of mite (*Tetra-*

nychus tiliarius) moving about. These webs often look more like a layer of varnish than anything else; but the reddish-coloured powder on them represents the inhabitants of this wonderful Liliputian city.

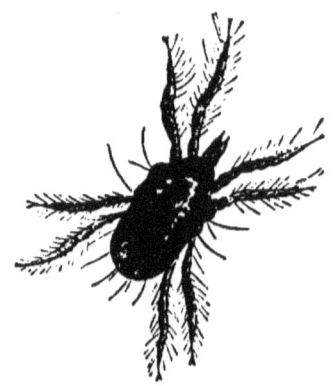

Fig. 218.—*Tetranychus telarius*. (The spot within the circle represents the natural size as nearly as possible.)

Fig. 219.—*Tetranychus tiliarius*.

The stone-mite (*Tetranychus lapidus*) was found in great numbers. It so happened that this had been a favourite object with Willie's father as a microscopical mount, on account of the singular beauty of its eggs. He told Willie that when he was a student at the Paris hospitals, he had frequently seen many of the stones in the promenades there covered with them. The eggs are white, although the mite is red.

But eventually, as regards land-mites, the old garden at the back of the house proved to be the best hunting-ground. This was fortunate, because

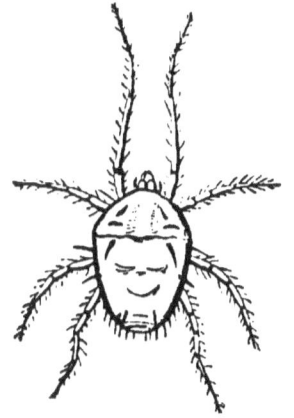

Fig. 221.—Eggs of stone-mite (*T. lapidus*).

Fig. 220.—*Tetranychus lapidus*.

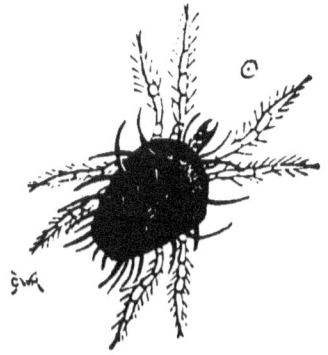

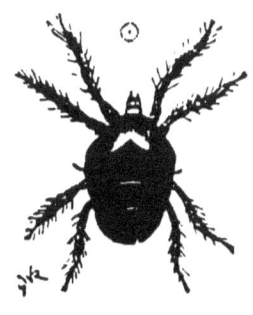

Fig. 222.—*Tetranychus ulmi*. Fig. 223.—*Tetranychus salicis*.

on wet days they could make a rush (after they had learned what to rush out for), and "collar" one or two different kinds. Of course, I don't know what the word "collar" means, but my young friends seemed to.

In this way they got the plum-tree mite (*T. prunicolor*), the elm-tree mite (*T. ulmi*), the willow mite (*T. salicis*), the poplar mite (*T. populi*), the guelder-rose mite (*T. viburni*), etc. In their rambles

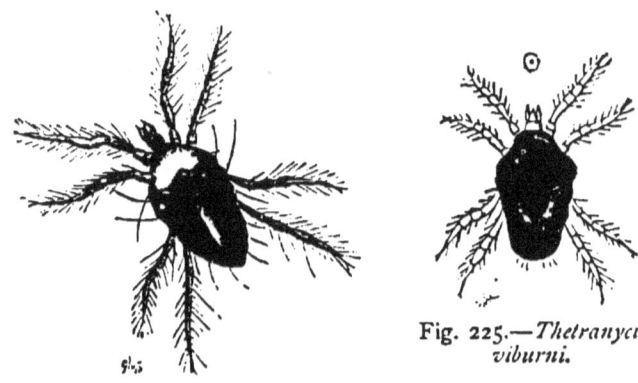

Fig. 224.—*Tetranychus populi.*

Fig. 225.—*Thetranycus viburni.*

they found that even the stinging-nettle—marvellously protected though it be—is attacked by a special species of mite (*T. urticæ*).

Do we really understand what is meant by the word "life"? If its importance is to be measured by the mass of flesh in which the subtle spirit is incarnated, then the Greenland whale and the elephant should be placed at the very highest summit of the zoological ladder. It can be imprisoned within, and direct the motions of, a microscopic body as easily and marvellously as those of a macroscopic body.

"As true, as perfect, in a hair as heart."

Life may even be "pill-boxed" within life, the animal within the vegetable, the vegetable within the animal—not *parasitically*, and therefore de-

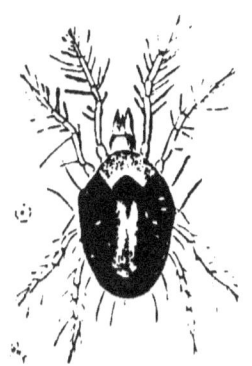

Fig. 226.—*Tetranychus urticæ*.

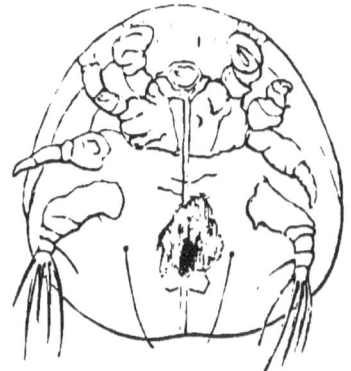

Fig. 227.—Mite from Gamasus of humble-bee.

structively, merely, but with mutual co-operation and advantage (*symbiosis*). This wonderful chain of life, of which Pope, with genuine poetic insight, said—

"In Nature's chain, whichever link you strike,
Tenth or ten thousandth, breaks the chain alike,"

and to which the great Darwin devoted a noble life for the purpose of making out its interdependence and absoluteness, and therefore justifying "the ways of God to man,"—loses none of its wonder, nor even of its mystery, as we know more of it. "The greater the circle of our knowledge, the greater the periphery of the external darkness." Lord Lytton's aphorism holds especially good of all genuine scientific research.

Swift was thought to be only using poetical licence when he said—

> "For bigger fleas have little fleas
> Upon their backs to bite 'em,
> And little fleas have lesser fleas,
> And so *ad infinitum*."

Here is a humble-bee, one of the very commonest of objects. Most of them are attacked on the under part of the body by minute, shiny, brown beetles called Gamasus. The belly of the bee sometimes swarms with them. Well, on this parasite is another—a mite (Fig. 227). Sometimes a Gamasus will have six or seven mites living upon it. They possess special kinds of claws for clinging to the Gamasus with, and a special kind of mouth.

But of all the various species of the family of mites, perhaps none are so beautiful or interesting as the water-mites. When our young naturalists set about mite-collecting, Willie remembered that in the volumes of *Science Gossip* for 1882, 1883, and 1884, there were some capital illustrated articles on the subject by Mr. C. F. George; so they had recourse to them to make out their finds. These water-mites belong to special groups known as Hygrobatidæ and Hydrachnidæ. They have generally from two to four eyes. When undergoing their insect-like changes, or *metamorphoses*,

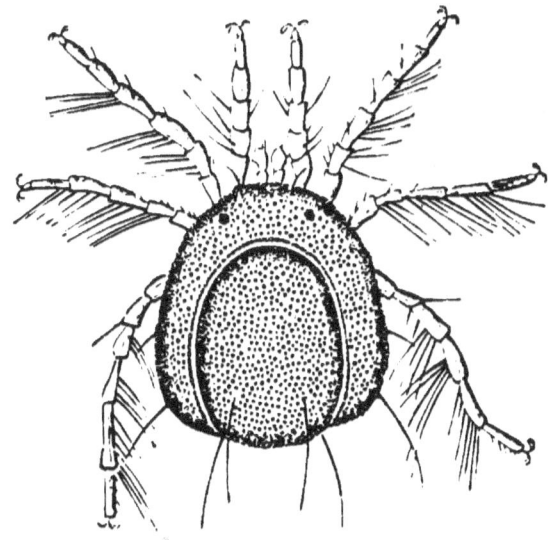

Fig. 228.—Female of *Arrenurus*.

a very significant fact occurs. Many naturalists now regard the different progressive stages through which an individual passes before it becomes adult, as more or less representative of the evolutionary changes through which the species itself has passed during the long period of time in which it has been developing.

For instance, when a water-mite is in its *larval* stage, it has only six legs. Therefore at that period, so far as the number of legs is concerned, it is an *insect*. In its adult stage it has eight legs. So the water-mites, in this respect, connect the two great classes of Insecta and Arachnida.

These two kinds of water-mites are separated into distinct divisions, one possessing two eyes only, the other four. The former go by the scientific name of Hygrobatidæ, and the latter of Hydrachnidæ.

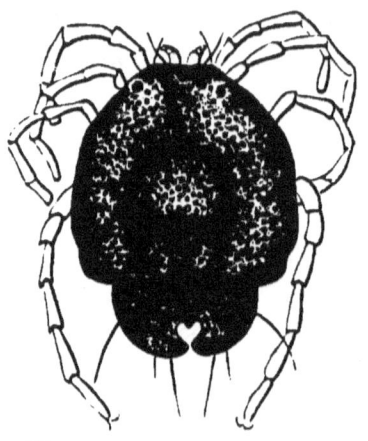

Fig. 229.—*Arrenurus perforatus*, male.

Our little party were successful in capturing several species of these creatures. Some seemed to prefer the clear, moving water, others the swampy or boggy places; some of them were of a green colour, like *Arrenurus viridis*, the males of which can be distinguished by their comparatively long tails. The genus Arrenurus includes several

British species, nearly all of which are brightly and beautifully coloured—green, blue, red, yellow,

Fig. 230.—Under side of *Arrenurus perforatus*, male.

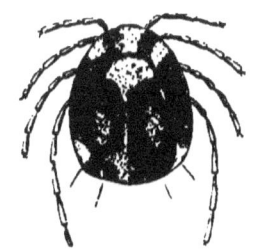

Fig. 231.—*Arrenurus buccinator*.

Fig. 232.—*Arrenurus buccinator* (under side).

etc. Nor do these colours fade after the objects have been killed and mounted for microscopical examination. Some are distinguished by having a hard or chitinous skin, others by possessing a soft one. The chitinous kind, however, is the more numerous. The hard plates fit almost like

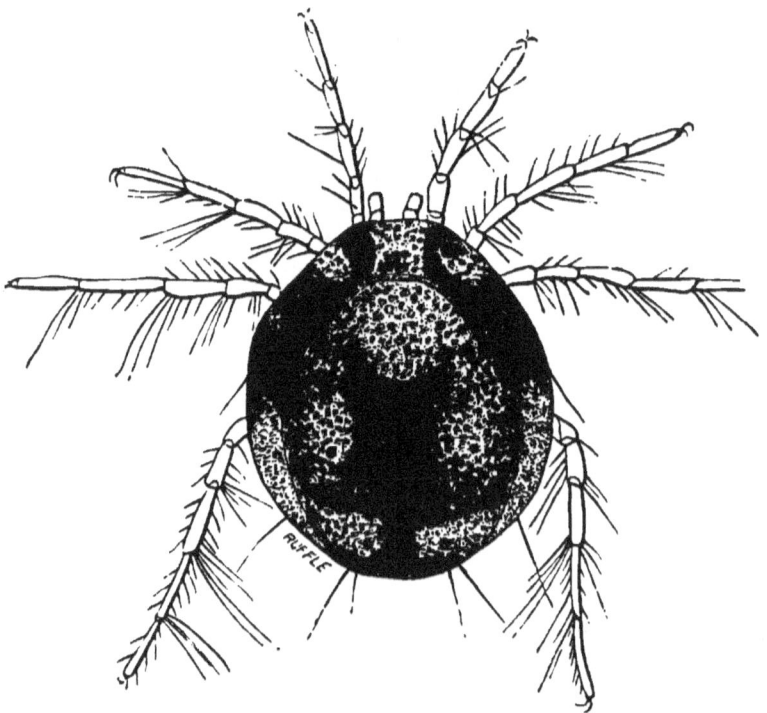

Fig. 233.—*Arrenurus ellipticus*, male (upper side).

those of the carapace, or shell, of a crab. Males and females of each kind are remarkable for their non-resemblance to each other. The eyes of nearly all are very beautiful objects when seen under a microscope.

Mr. George tells us that when they are confined in a glass vessel of water, the females lay their eggs on the glass. The eggs are generally of a pinkish colour, surrounded with a whitish opaque

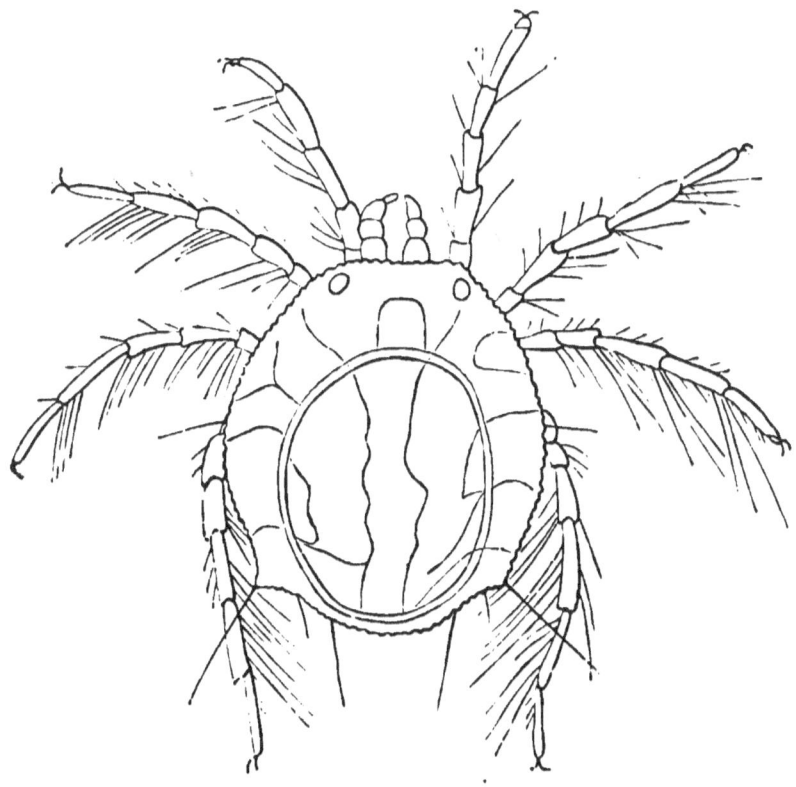

Fig. 234.—*Arrenurus tricuspidator*, male.

substance, which seems to be the material used in cementing the eggs to the glass. When the eggs hatch, a minute larva is produced, possessing six legs. From their appearance, Mr. George thinks they must be parasitic on some other kind creature; but he never could find out which. So here is another riddle still left unanswered for such students as my young friends.

Among the captures were *Arrenurus sinuator*, whose short tail is of a bright yellow colour,

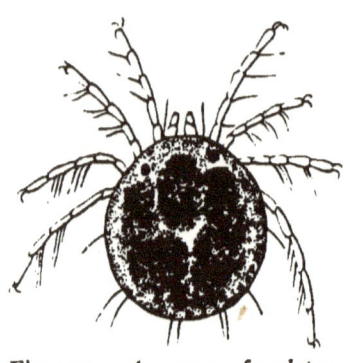

Fig. 235.—*Arrenurus frondator*, female (upper side).

and the part where the body and tail are joined together, a beautiful blue; *A. albator*, of a light

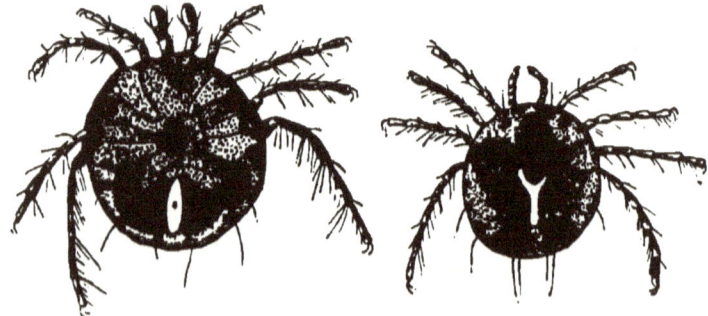

Fig. 236.—*Arrenurus rutilator*.   Fig. 237.—*Arrenurus rutilator*, female (upper side).

body colour, and having a differently shaped tail; *A. crassicaudatus; A. perforatus*, one of the most

A NEW HUNTING-GROUND. 175

Fig. 238.—*Arrenurus integrator.*

Fig. 239.—*Arrenurus truncatellus.*

beautiful of all our British water-mites; *A. buccinator; A. ellipticus; A. tricuspidator*, etc.

Fig. 240.—*Arrenurus globator*.

Fig. 241.—*Arrenurus globator*, male.

The female water-mites are not only more numerous than the males, but of larger size.

Among the soft skinned water-mites, the commonest are *A. frondator*, *A. rutilator*, etc., all very small, and of a globular shape, but hardly less brightly coloured than those above mentioned.

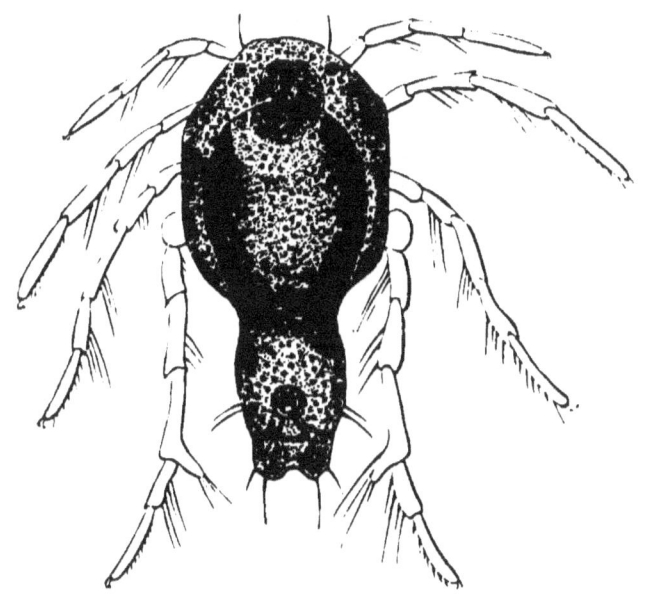

Fig. 242.—*Arrenurus buccinator*.

Arrenurus tricuspidator, *A. integrator*, and *A. truncatellus* are less common forms; the latter is an exquisitely lovely object, green, with vermilion eyes. *A. integrator* is of a lovely blue, and *A. tricuspidator* of an equally attractive red. All the puncturings and other markings came out

178   THE PLAYTIME NATURALIST.

splendidly when these lovely creatures were put under the microscope, and examined as opaque objects. Jack's father was quite interested by

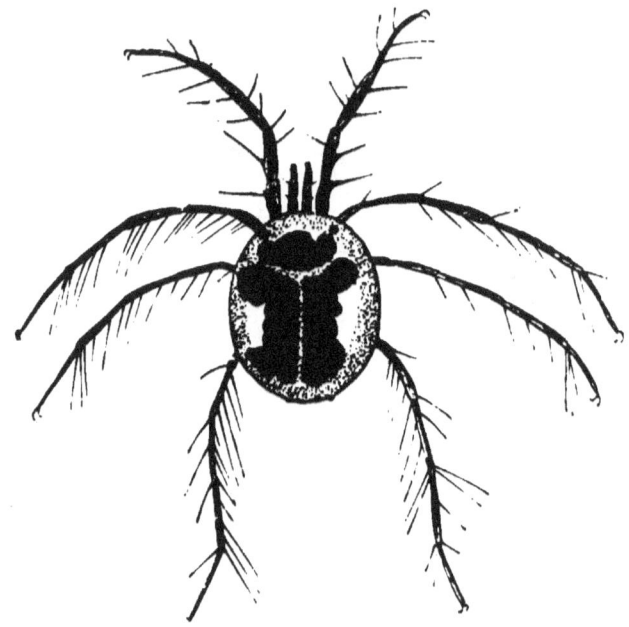

Fig. 243.—*Arrenurus atax.*

their beauty; and as for Jack's sister—why, she made a mental vow she would never marry a man who was not a naturalist!

CHAPTER IX.

TOADS, FROGS, NEWTS, AND REPTILES.

I FORGOT to mention that, earlier in the summer, the two lads had done the usual preliminary aquarium-keeping. That is to say, they had got up a structure with glass sides which leaked horribly, and had crammed it with all sorts of water-weeds and water-creatures—molluscs, fishes, frogs, newts, water-beetles, etc. There was terrible murder and massacre for a day or two; splendid eating and drinking for a few; and suffering and dying for the many. Nature will forgive almost any crime except that of overcrowding! Her punishments for this offence are unappealable.

The eager young naturalists soon recognized this important fact. They had put into their leaky old tank all the things they had found, from humane motives, not cruel ones. There was, of course, the desire of possession, and the joy of conquest. But neither of the lads would have caused the slightest suffering if he could have helped it. And as they fished out the dead bodies of their previous day's hunting, both of them said, "Poor beggars!" That was their Burial Service. Then, like brave and sensible lads, they recognized the fact that failure is only the stepping-stone to success—that a boy (as well as a man) may indeed learn more from a single failure than a single success; that is, if he has got any grit in him. If he hasn't, why, it doesn't matter much, either way.

The microscope had come to the ardent young fellows like a quiet revelation. It had made every living fact worth observing, all the more observable. Consequently, after the *coup d'état* of their aquarium, the aquatic government of that colony had settled down into a little better order. There were fewer living creatures in it, and consequently fewer rows. Some people wonder why Europe should be practically an "armed camp." They forget that its population has doubled, and, therefore, we have double as many people to quarrel among themselves now than there were in the days of Napoleon Bonaparte. It is not countries which go to war; it is nations —that is, populations. Germany and France do not hate each other, or despise each other, more than the dwellers in the East and West Ends of overpopulated London. Overcrowding means murmurings, grumblings, wrestlings. Overcrowding is punished by Nature after her own stern manner. Rachel weeps for her children in vain, and refuses to be comforted, when the germs of typhoid and cholera attack overcrowded and therefore uncleanly cities. Farmers grumble, and pray, and starve, when they overcrowd their corn-fields; and moulds, rusts, and mildews visit them in consequence.

There is just another point worth mentioning in this sermon of mine. The higher the zoological rank of organisms, the more their requirements.

But I am not going to stop my narrative for a series of copy-book moralities; so the end of my sermon has come, and its application, which is—if you rig up an aquarium, don't put too much into it. When Nature first rigged up our planet with life, she began with small and feeble things. That enabled her to get on, to add to her stock, to introduce more highly organized breeds, until at length earth became the Chief Nursery for Heaven.

My young friends had collected "a lot of spawn." That comprehensive boyish term included the eggs of newts, frogs, and toads. Of course, people will persist in calling these objects *reptiles*, whereas they are actually more nearly related to fishes than to reptiles. They are "nasty slimy things" to most people; to others who are in the secret, they are the degenerate descendants of a race of primary vertebrates which once held the same relative position among the existing tribes of the earth as you and I do to-day.

There were once English frogs as big as Highland oxen, and marine newts as large as ordinary crocodiles. In Cheshire (about Storeton and Liverpool generally) you find their footprints, bigger than the impressions made by human hands. The huge Japanese salamander in the Zoological Gardens (three feet long) is the last giant representative of a race which flourished at its best many millions of years ago. No wonder, instinctively feeling the superiority of its genealogical position, that it snaps so fiercely at the umbrellas and sticks which "'Arry and 'Arriet" poke into its face, when it is trying to get a *siesta*.

Take the spawn of toads, frogs, and newts, for instance. There are no commoner objects anywhere. This so-called "spawn" is merely the eggs of those creatures. All true amphibians resort to the water to deposit their eggs therein; whereas reptiles avoid the water for that purpose. Even such thoroughly aquatic reptiles as turtles will swim hundreds of miles to deposit their eggs on the land. And then, the young of all amphibians live and swim in the water, and are possessed of special swimming and breathing organs for the purpose, except in those easily explained exceptions (such as the European salamander) in which the larval stages are *accelerated* to avoid such a necessity.

The gathering together of frogs and toads in the ponds and marshes in the early summer, and their gratified calls to each other ("croaking," the uninitiated call such noises), is heard all over the world. It is as common in Australia as in England —even commoner. The wide geographical distribution of these creatures is a proof of their high geological antiquity.

When my young friends first went out to seek the spawn of these much-despised but thoroughly harmless creatures, they half shared the usual dislike universally manifested for them. Whereas it is a pleasure to find and to handle birds' eggs, it is with much overcoming of prejudice only that people can be brought to handle the eggs of toads and frogs. Nevertheless, the examination and observation of the latter objects are even more interesting than birds' eggs. In the latter, the opaque egg-shell hides all observation of the marvellous transformations which take place whilst the germ-spot of the yolk is converted into a bright, feathered, attractive bird. In amphi-

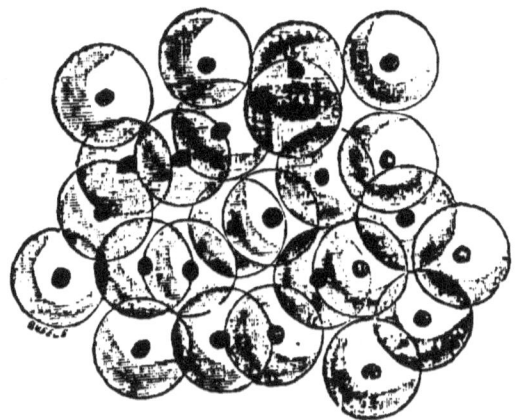

Fig. 244.—Frog-spawn *in situ*.

bian eggs, with low microscopic magnification, you witness every stage in the wonderful series of changes, until the larvæ, or " tadpoles," are hatched out. Even then the half-transparent tail and gills enable you plainly to see the circulation of blood.

The young collectors were not long before they could distinguish between the spawn of toads and frogs. That of the former is in long strings, like so many necklaces; that of the latter, in dense irregular masses. The newts take more trouble, and wrap each individual egg in the leaf of a submerged aquatic plant, such as the leaves of the star-wort (*Callitriche verna*), for instance.

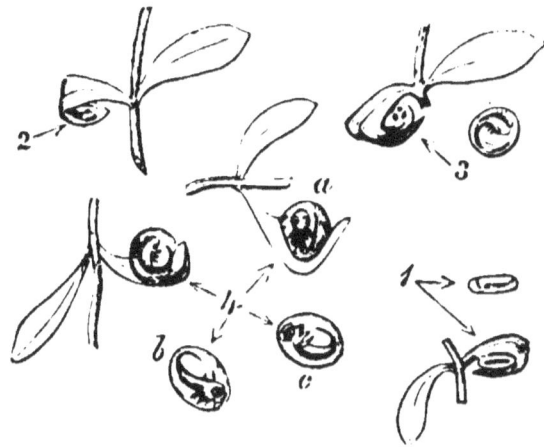

Fig. 245.—Single eggs of newt wrapped in leaves, showing development.

The boys kept the frog-spawn in a shallow vessel, which they covered with glass, to exclude the dust. Into this vessel was placed a supply of small aquatic plants, so that when the tadpoles hatched out they would find a plentiful supply of infusoria on which to exist after absorbing and devouring the gelatinous masses of their eggs. There are no better scavengers, removers of decaying material, in the world than tadpoles. This glass vessel was placed in a window having a north-east aspect. The spring was late this year, so the observations on the spawn were carried on right into April. The eggs, as a rule, do not

 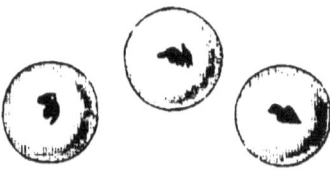

Fig. 246.—Small ova of frog.   Fig. 247.—Cleavage or segmentation of eggs (sixth day).

change much until the fifth day, although their integuments thicken and get rather tough. Then a segmentation is observable, which becomes opaque and more distinct every day. On the ninth day a striking change takes place, and the tadpole

Fig. 248.—Ditto, another stage.   Fig. 249.—Ditto, advanced stage.

is roughed out, so to speak. On the fourteenth day the embryo has the power of self-movement, and you see it bringing its head and tail together with a jerk. On the fifteenth and sixteenth days the juvenile tadpoles escape from the eggs, but they have a good deal of wriggling to do before they clear themselves of their albuminous investment. Afterwards they set to work and devour it!

Fig. 250.—A, Segmentation of egg, fourteenth day; B, ditto, enlarged.

The length of the young frog-tadpole, when it escapes from the egg, is not quite half an inch. The gills grow very rapidly, and it is a pretty sight to behold the blood slowly but regularly circulating through them. The best way of observing them is to rig up a zoophyte-trough by taking two strips of glass, and fastening on

Fig. 251.—Development of tadpole seventeen days after laying the egg.

one (A) an india-rubber ring (D), deep enough to hold sufficient water in which to put the tadpole, or any other similar aquatic object you want to observe in the living state. Then place a strip of glass over the top of the improvised cell, and fasten the two strips together by means of the ordinary thin india-rubber bands (B, B) used for paper rolls. Two small wooden wedges (C, C) keep

Fig. 252.—Tadpoles sixteen days after laying the egg (natural size and enlarged).

the glass strips so well gripped that the cell will be quite water-tight.

It is very interesting to observe the changes in the tadpoles. As the warm summer months develop them, you see them changing, not only their shape, but their habits as well. They

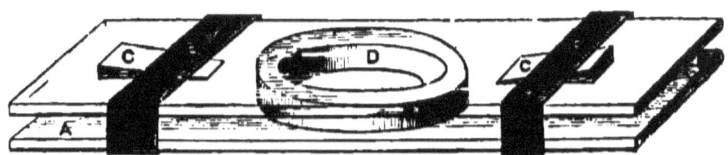

Fig. 253.—Improvised live-box.

frequently come to the surface of the water to breathe; they swim less with their tails, and use their developing limbs for that purpose instead.

In the tadpoles of toads you see the action of the fore limbs confined beneath the translucent

skin, moving synchronously with the after-limbs.
Toad-tadpoles behave very similarly to those of

Fig. 254.—First frog-stage (magnified).

Fig. 255.—Second frog-stage (magnified).

the frog. At the end of the long series of changes

the tadpole rather suddenly assumes the adult shape. The tail and gills are absorbed, and the body shrinks, as it were, into the shape of a frog or toad, as the case may be. About 115 to 120 days from the deposition of the spawn are required to attain the adult and natural form.

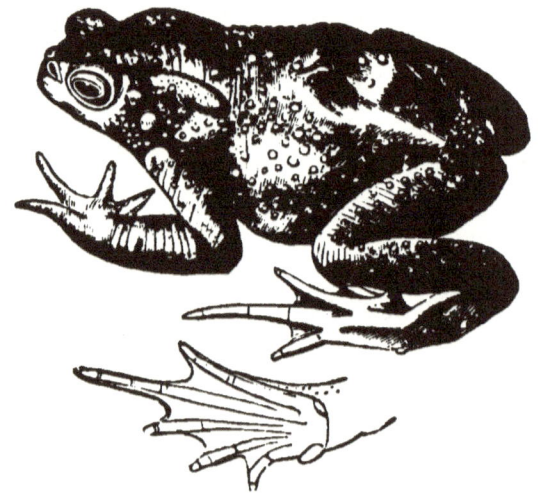

Fig. 256.—The common toad.

Newts, as everybody knows, never let go their tails, although they lose their gills; some retain the latter for a year or two after leaving the larval state. The mature newt, therefore, represents the tadpole state of a frog or toad, and the latter may be regarded as an advance on the former.

Those really beautiful objects, the crested newts

(*Triton cristatus*), common in every pond, were procured, male and female. Everybody was delighted with what some of them had before despised, when they saw the brightly coloured males moving about in the large bell-glass so gracefully —their dorsal crests waving like a shirt-frill, and their brilliant eyes looking so intelligently at the spectators. I wonder what the newts thought of the laughing and joking band of young philosophers on the other side of the glass!

The smooth newt (*Lissotriton punctatus*) is

Fig. 257.—Female of smooth newt.

another of our British amphibians. It is most interesting to watch the female depositing her eggs. She does not leave them in strings and masses, like the toads and frogs, but wraps each ovum up separately in the living leaf of an aquatic plant (see Fig. 245), as if it were the most valuable little parcel in the whole world, which it doubtless is to the careful and anxious mother. These eggs require from fourteen to sixteen days to hatch. The dainty little tadpoles which emerge are much more transparent-looking than those of frogs and toads. Indeed, they are so very transparent at first that you have to look keenly to see them; so I suppose their transparency must be more or less protective. Their fore feet develop within four or five days of hatching, and then the creatures are about half an inch long.

The boys had what they called "some rattling

Fig. 253.—Male of smooth newt.

good fun" at home one day. There is no fun a boy likes more than *teasing*, and especially in teasing those he loves most; and whom he would defend and fight for, tooth and nail, if any other fellow dared to tease them.

Now, the great art of teasing is to frighten your sisters and cousins with something you are not frightened at—at least, not now. You must not tell them that once you shuddered as much when you touched a frog or a toad as they do now—that would spoil the fun, and lower your own dignity. Nor must you in any way let them know that even now you regard snakes and vipers with the same fear and trembling that your sisters do frogs and toads.

All this "teasing," however, is very human; and I fully believe sisters and cousins like it, although they call the inflictors "horrid old things." Life would be a very dull affair if it were not for a little breeze, just as the atmosphere would get stagnant were it not for storms. Even love-birds get up sham quarrels in lieu of real ones. You see, they are like all the rest of us—are "obliged to keep up appearances."

The lads had taken a couple of specimens of the natterjack toad (*Bufo calamita*)—a much rarer species than the common toad, and usually restricted to swampy districts not far from the sea. A yellowish-brown sort of toad, remarkable for the yellow line running down the middle of its back, and the black bands on its legs. It runs about almost like a mouse, and has quite a different style to that of its commoner relative.

Well, the lads let out a couple of these natterjacks in the room, and you can imagine the commotion they created. If they had crawled like ordinary toads, that would have been bad enough; but to see them *running* was what no toad had ever done before. So there were shrieks, and jumping on chairs, and cries of "Horrid things!" until the mischief-makers captured their prey in a corner, which they were easily able to

Fig. 259.—Natterjack toad (*Bufo calamita*).

do, being led there by the disagreeable smell of the creatures.

There was a fine patch of heath-land not far from the house—just the very place to look for slow-worms, lizards, and vipers. And hard by was a damper tract, where ordinary snakes were known to frequent. Thither the young adventurers marched, with bottles to contain their prey, and some benzine and spirits of wine. There was

TOADS, FROGS, NEWTS, AND REPTILES. 195

Fig. 260.—Group of British lizards. *a*, Viparous lizard ; *b*, sand-lizard ; *c*, blind or slow-worm,

a sort of danger mixed up with this excursion, which somewhat spiced it—the dangerous adders, or vipers. Of course, these creatures are not to be despised. They really can inflict a very painful and a very dangerous bite, and there is no doubt whatever that occasionally such bites have ended fatally. Nevertheless, I am of opinion that the

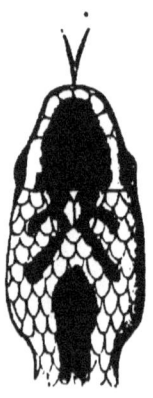 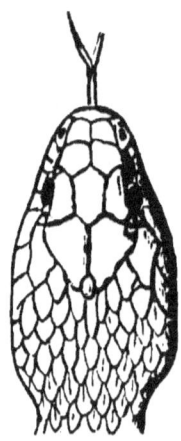

Fig. 261.—Head of viper.   Fig. 262.—Head of common snake.

viper is a very much over-dreaded reptile. I am very certain it is much more afraid of human beings than we need be afraid of it. Still, for all that, I never approach a sleeping viper except with cautious respect.

The viper is a capital illustration of the benefits to be derived from " putting on a lot of side." It looks a good deal more dangerous than it really is, and threatens a great deal more than it is capable of performing. It is almost ludicrous to see how helpless a viper is when you get hold of it by the end of its tail. It wriggles and twists and tries to get at your hand; but it can no more do so than if it were a tortoise. I have sometimes thought—in fun, of course—that the "death's head and cross-bones" sort of pattern on the viper's head, which distinguishes it from the head

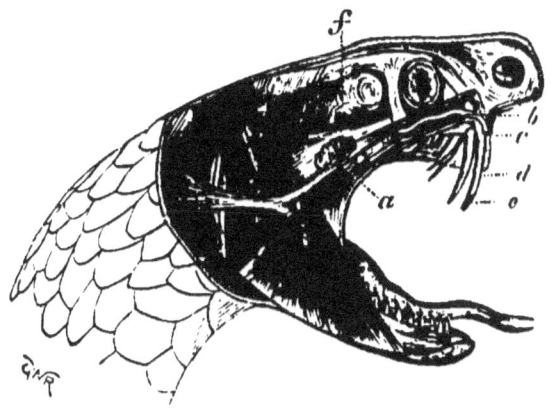

Fig. 263.—Section of head of viper. *a*, Poison-glands; *d*, fangs.

of the common snake, is only a sort of "Pirate of Penzance" swagger about the "black flag," etc. The common snake is content with a ring round its neck—hence its zoological name of *Natrix torquatus.*

Well, two fine vipers were captured, Jack, who

had had a good deal of experience in such matters, showing the others how to secure a sleeping viper with a forked stick, which held the head down immovably. Then he acted as dentist, and removed the two long fangs from the upper jaw of the creature with his penknife. These are hollow, and communicate with the poison-glands, so that when the reptile bites the poison flows into the wound. It is this structure which makes poisonous snakes so dangerous. When Jack had removed the fangs, the vipers were quite harmless, and he then let them do what they liked—which they couldn't.

The viper is not an affectionate pet, even when its poison-fangs have been removed. Now, the common snake can be educated and trained into one—to come at call, and to behave decently. It is a splendid and graceful swimmer, and loves its bath quite as much as the boys who fling it in like to see it swim in such elegant undulations.

Two or three species of lizards were caught among the dry roots of the half-dead furze-bushes and those of the heather. These lizards are not easy things to capture. They are out of sight in the twinkling of an eye—except the tips of their tails, which disappear so quickly that you are not quite sure it was not a viper which vanished. One species, the common brown lizard (*Zootoca vivipara*), is a pretty little fellow, with a glinty brown back and pale orange-coloured belly, which is easily domesticated in a "vivarium." The female retains her eggs in her body, where they hatch, and the young issue forth alive—hence the specific zoological name.

The sand-lizard is distinguished from the above both by its different habitat, and different markings. Slow-worms are among the most interesting and harmless of our British reptiles, although I have seen ladies who were not afraid of their husbands, very much afraid of slow-worms. If the doctrine of *metempsychosis* be true, that is what the husbands of such wives ought to turn into.

Then again, the slow-worm is a genuine British creature. It embodies in its name genuine British prejudices and blunders. In the first place, it is not very *slow*; in the second it is not *blind* (far from it); and in the third, it is not a *worm*. Stupid people sometimes inquire why naturalists give such long and accurate names to things; the above is an illustration of the accuracy of "popular names." It reminds one of that festive confection without which no British Christmas would be perfect—the plum-pudding. The fruit which bestows its name upon it is a small *grape*, which when dried is called a *currant*, and this currant proceeds to be a *plum* when it is put in a pudding! It is, perhaps, this bungling of words which makes the English plum-pudding so dear to the English people.

The slow-worm is in reality a *footless lizard;* neither a worm nor a snake. Its small but pretty eyes can soon be found, if a person is not afraid of it. And there is no earthly reason why anybody should, for even if a slow-worm was wicked enough to try to bite you, it couldn't draw blood; therefore it is more harmless than a flea.

The boys lost several fine and beautifully purple-glinted slow-worms through getting hold of them by their tails. The slow-worms are always willing to sacrifice the fag-ends of their tails to save their bodies. So the lads got a few joints, that was all. But, anyhow, they learned that slow-worms have been able to turn even *catalepsy* to defensive purposes, and that they can stiffen their tails sufficiently to break off like a bit of dry gingerbread, whilst the living body so protected creeps out of sight, and also out of danger!

CHAPTER X.

SMALL FRY.

MEANTIME, during wet days and in the now extending evenings, the microscope had been more used, both for amusement and investigation. Then it was that surface-dredging of the streams and ponds commenced in earnest. The boys learned how to make delicate muslin bag-nets, and not to trail them too quickly through the water, for fear of bruising the fragile things they wished to capture and examine. They learned, also, how to invert the muslin net, so as to make the inside the outside, and then to dip it repeatedly in their wide-mouthed pickle-bottles, when, of course, all the contents were safely washed off. It was real fun to see how the bottle was scanned with the pocket-lens after every such dip and wash. The eye soon learns to detect the presence of a new-comer, even in a pickle-bottle. Its ways, antics, movements, all make it out as "something new." So there remains the anticipated pleasure of further and better observation of the capture in the "zoophyte-trough" or the "live-cell" with the microscope in the evening.

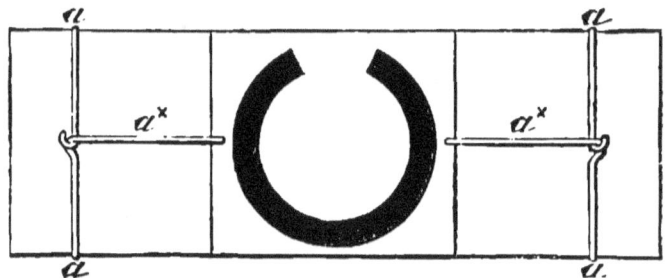

Fig. 264.—An improvised zoophyte-trough.

The accompanying sketch will give a good idea of how to rig up a temporary "live-cell" for the employment of powers ranging up to and including a quarter-inch. You get a glass strip, and one

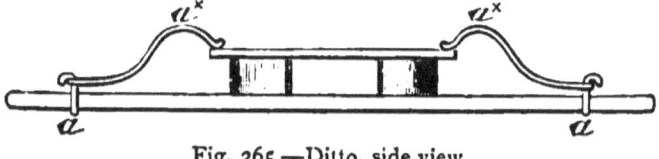

Fig. 265.—Ditto, side view.

of the thin round glass rings sold by all dealers in microscopical materials. Then snip a bit out of the ring (indeed, properly prepared snipped rings can be got at about a shilling the quarter-ounce, which will contain hundreds). Place this snipped ring on the strip, then a drop or two of the water containing objects to be examined, and over all a cover-glass. If the snipped cell is not too thick,

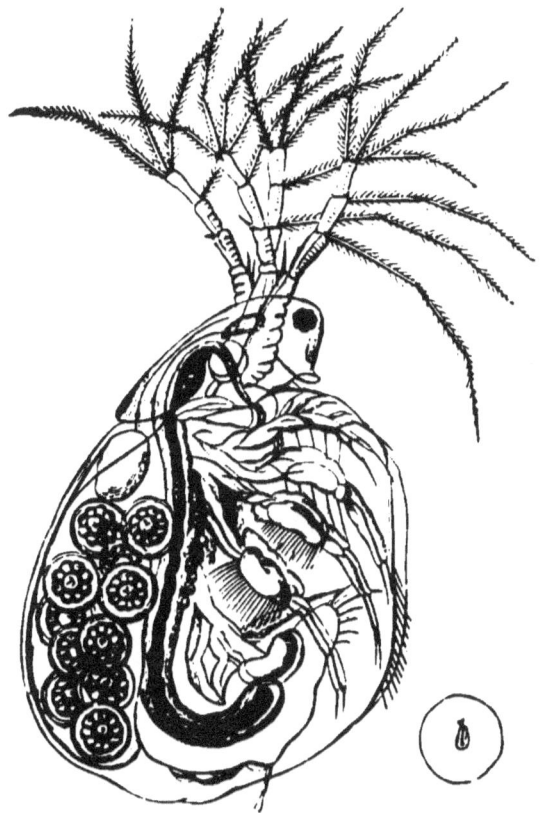

Fig. 266.—Water-flea (*Daphnia vetula*), female. The natural size is shown by the figure in the circle.

the whole will hold together by the cohesiveness caused by the moisture; if it is rather thick, you can easily clinch them by a bit of wire (*a*).

The first things obtained in the pond were the little entomostracans, better known as "water-

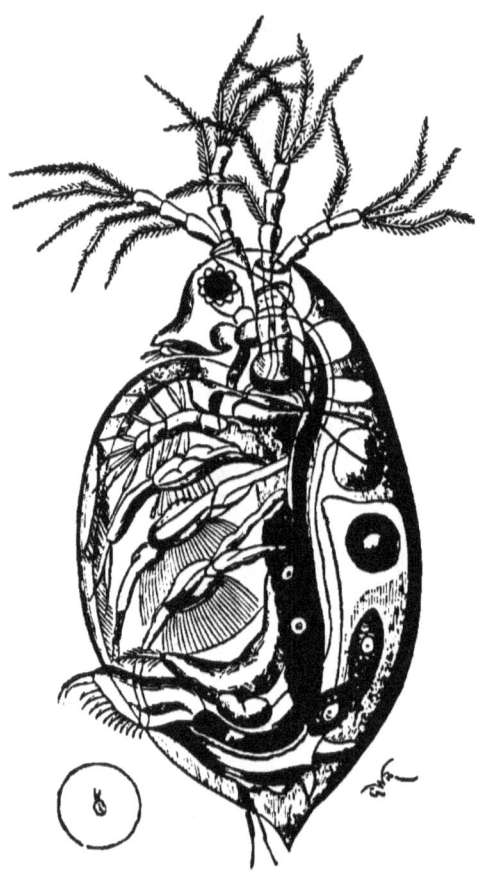

Fig. 267.—*Daphnia pulex*.

fleas," which abound in all still waters—sometimes to such an extent as to give the water a dark and cloudy colour. If the collecting bottle containing them be held up to the light, you can perceive them with the naked eye, fussing and jerking about in almost ridiculous movements, in company with their near relatives, the cyclops. These common "water-fleas" are among the

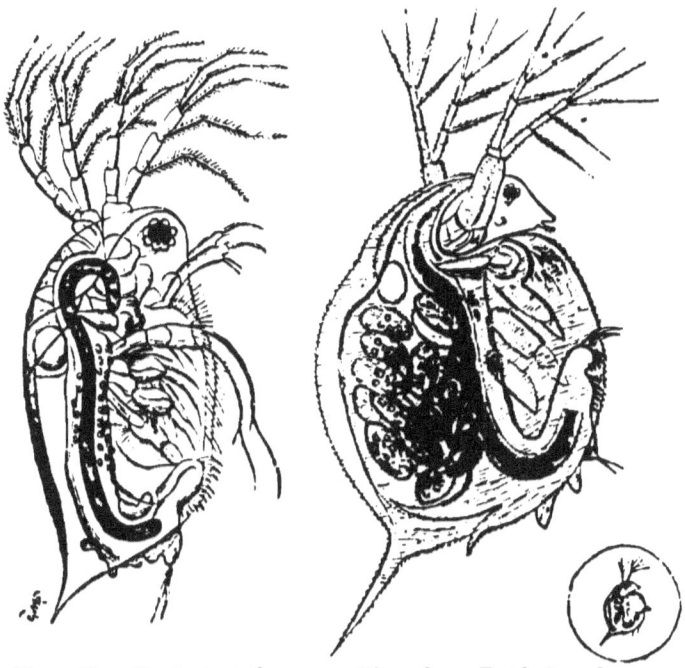

Fig. 268.—*Daphnia pulex*, male.

Fig. 269.—*Daphnia pulex*, female.

oldest existing kinds of creatures in the world. Their genealogy can be traced backwards in an unbroken line, through every geological period beyond the Carboniferous. They were as common in the standing pools surrounded by the strange vegetation of which our coal was formed as they are now in an English pond or lake. The covering of these lowly organized crustaceans is chiefly *chitinous*—a substance which is very durable. This covering is like a swan-mussel, fastened on one side and free on the other; and it is usually so thin and transparent that all the bodily organs are visible through it. The gills

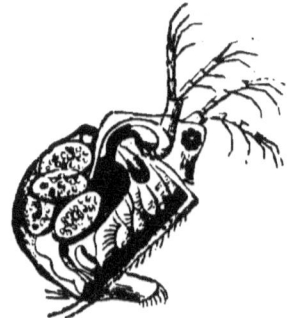

Fig. 270.—*Daphnia reticulata*, male.   Fig. 271.—*Daphnia mucronata*, female.

are worked by the movements of the feet, and are exquisitely graceful objects. When these creatures die, their thin shells fall down to the bottom among the mud.

In many of the black shales overlying our English coal-seams (all of which were once fine, soft black mud), you may see the surfaces strewn with myriads of the fossil shells of the tiny water-fleas which lived millions of years ago.

There are several species met with in our lakes and ponds. The male and female differ from each other sometimes in a very marked manner. The eyes of such genera as Daphnia are very remarkable, and very beautiful. The structure of the mouth is also peculiar. The female can be immediately detected by the eggs seen within the semi-transparent carapace.

Willie was successful in dissecting and mounting one of the compound eyes of a daphnia, so as to show the lenses, the optic nerves, etc., the result being as is shown in the figure.

There is quite a rage just now for collecting the lower kinds of fresh-water crustaceans in our lakes and reservoirs, as well as in large ponds. They are dredged slowly, with a fine muslin surface-net, the inside surface of which is carefully washed when hauled in. It is surprising to see the number and the exquisite beauty of the perfectly transparent and glass-like minute crustaceans obtained in this manner.

But of all the interesting and even amusing creatures of this extensive class give me the *cyclops*. It is a standing favourite among microscopists. Here is the female, with her highly sensitive antennæ, and her single large round cluster of compound eyes right in front of what

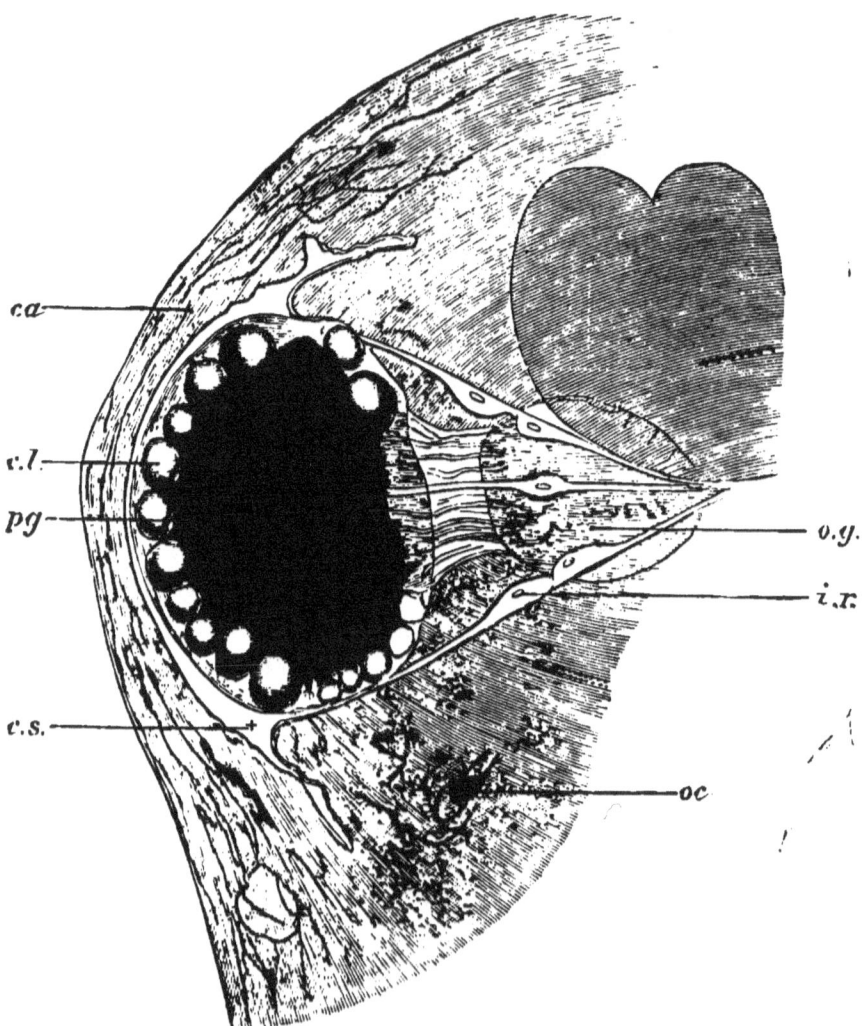

Fig. 272.—Eyes of water-flea. *ca*, Functional cornea ; *pg*, pigment; *og*, optic ganglion ; *cl*, lenses ; *oc*, simple eye ; *ir*, inferior muscle ; *cs* represents aqueous humour.

one is inclined to call her forehead. Behind her she trails two crowded egg-bags, one on each side resembling in shape the long silk purses our grandfathers used to carry. It is on account of the round cluster of eyes that these creatures have been called *cyclops*, after the fabulous one-eyed monster of Greek mythology.

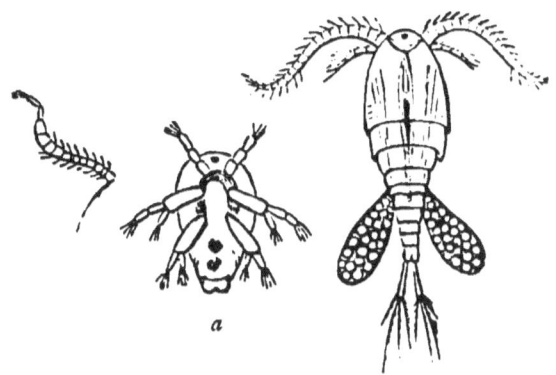

Fig. 273.—*Cyclops quadricornis.* a, Young.

The recently hatched cyclops is a comical little fellow. At first it has only three pairs of legs, and it moves about with a series of cranky jerks, first on one side and then the other, as if aware it has not yet got all the legs which properly belong to its class. If its motions had been specially designed to afford laughter, they could not be more successful. You may be sure my young friends used to make the most of this pantomime.

After a short time of feeding, however, the young cyclops moulted, and assumed another shape. A few more months, and it resembled its parent, male or female, as the case may be. The male can always be distinguished by the absence of the two egg-bags.

What tribes of creatures carry on their lives in these quiet ponds! Thousands of millions of ardently competing, voiceless creatures are thrown into each other's company. Our over-populated European cities are as barren places in comparison with the closely packed fauna of a single standing pool.'

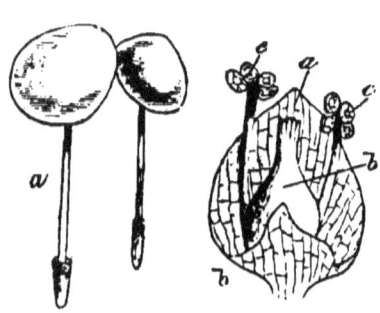

Fig. 274.—Duck-weed (*Lemna minor*).

Even the "green scum" which mantles its surface has a special character and history of its own. Look at these brilliant green duck-weeds (Lemna), for instance. They are real flowering-plants, although they are so small, and all the mineral matter they require is derived by the absorbent action of that comparatively long, thread-like root hanging down, and which is provided with a spongy tip for the purpose. Similar duck-weeds were covering the surfaces of extinct American lakes before the *Eocene* period, for they have been fossilized in the Dacotah beds of that country, formed in the long interval between the chalk strata and the London clay. Consequently, this

Fig. 275.—*Hydra viridis*, attached to duck-weed rootlets.

common class of the simplest structured of all flowering-plants must have been continuously in existence for several millions of years at the least.

The duck-weed root-threads are capital conveniences for many kinds of fresh-water microscopic animals. You will find colonies of Vorticella attached to them, or one or two species of hydra hanging from them. The latter are among the most interesting of all fresh-water objects; perhaps they represent the ancestral types of those common seaside objects, the corallines, sea-firs, etc., and even some of the branched limey corals of tropical regions.

The commonest species is *Hydra vulgaris*. Another, *Hydra viridis*, wants some looking for, as its green colour conceals it from view among the

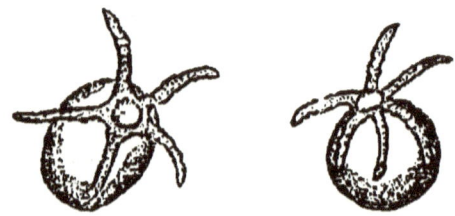

Fig. 276.—First stage in development of hydra.

water-weeds. It varies from one-eighth to a quarter of an inch in length. Perhaps you may get a compound specimen, as shown in Fig. 276, where the young hydræ have attained the size when they become detached from the parent body; or the parent body will be roughened with wart-like prominences, and even knobs. You may even be fortunate enough to catch one of these knobs budding with four tentacles, and watch it thenceforward until it sets up an independent

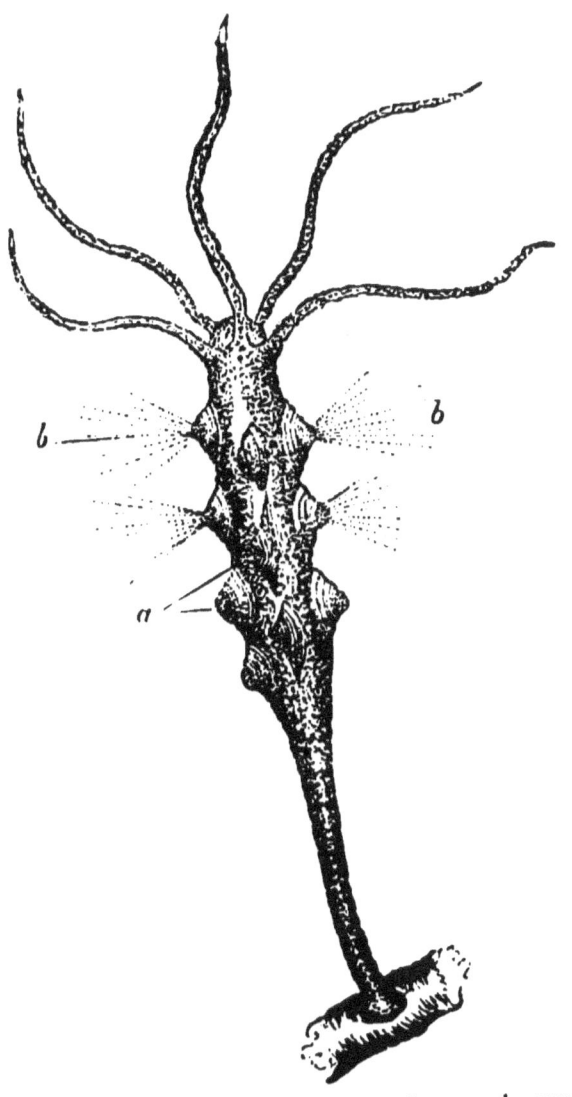

Fig. 277.—Hydra (magnified), showing prominences.
a, b. Eruptions.

existence for itself. Even then your observations are not over. You behold the number of tentacles increasing; you even witness the strange and almost magical manner in which the hydra—a

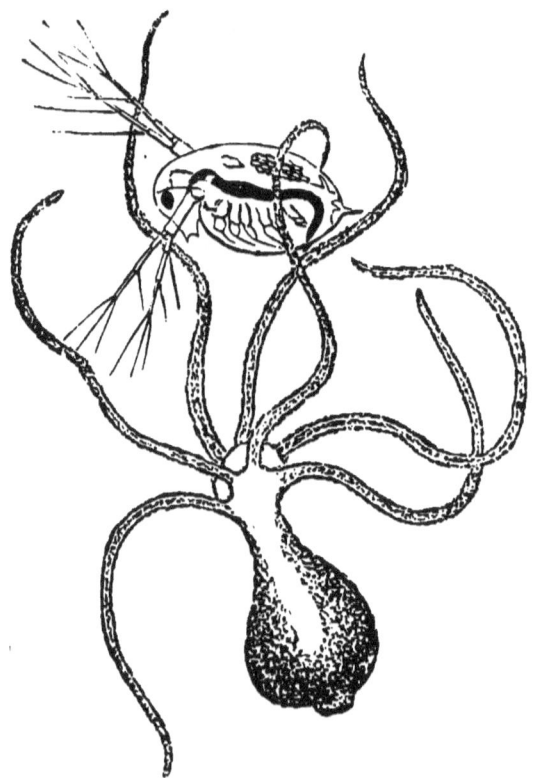

Fig. 278.—Hydra attacking a water-flea.

creature of lower organization — benumbs and secures such complexer and nimbler creatures as water-fleas (Fig. 278). The secret of this benumbing power is shown in Fig. 279, where the arrow-headed stings, or "urticating threads," are seen. These stinging threads are like those possessed by near relatives of the hydras, the stinging marine jelly-fishes.

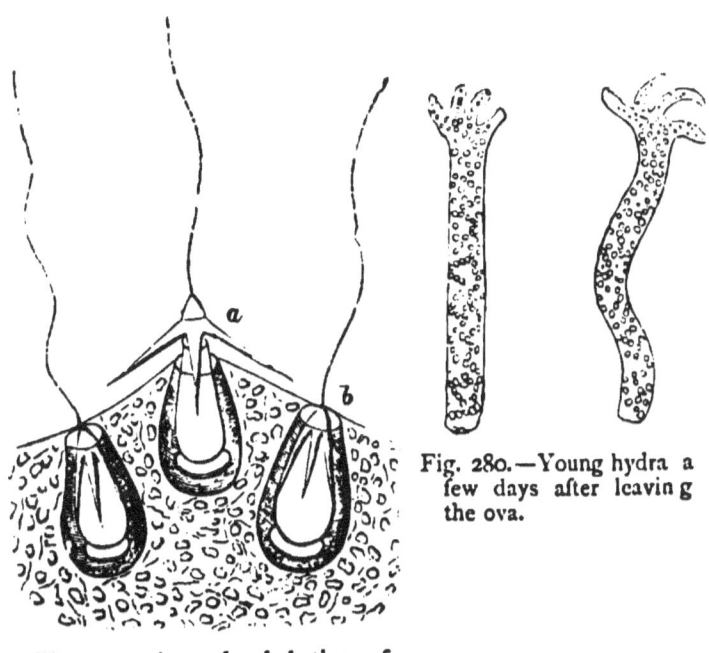

Fig. 280.—Young hydra a few days after leaving the ova.

Fig. 279.—Arrow-headed stings of hydra. *a*, Expanded; *b*, at rest (highly magnified).

Hydræ also reproduce their kind by means of eggs, which, when fertilized by spermatozoa, sink to the bottom of the pond. There they hatch, and the first appearance of a young hydra hatched from the egg is shown in Fig. 280.

It was usual during their dredging excursions to put the "best finds" in special bottles. The lads always took care, if there was anything attached to a leaf or stem, to snip the latter off without disturbing the creature, and drop it into the collecting-bottle. It was in this way they found that exceedingly lovely and not uncommon object, the fresh-water polyzoon Plumatella. This colony of relatively highly organized animals bear the same relation to the sea-mats (Flustra, Membranipora, etc.), of our seaside that the hydra does to the sea-firs (Sertularia). Their name of *polyzoa*, or "many-creatured," is in allusion to their habit of living together in colonies.

The Plumatella is not the only group of its kind. There are also the equally lovely genera, Fredericella, Cristatella, Paludicella, etc. All of them possess nervous structures and an elaborate and specialized mouth-apparatus for creating currents in the water—producing microscopical whirlpools, in short—the centre of which leads into the mouth, and sweeps all the living prey directly into it. Fig. 281 shows a group of several of these associated animals, some with the highly elaborated tentacles thrust forth, others as they appear when withdrawn into the body-

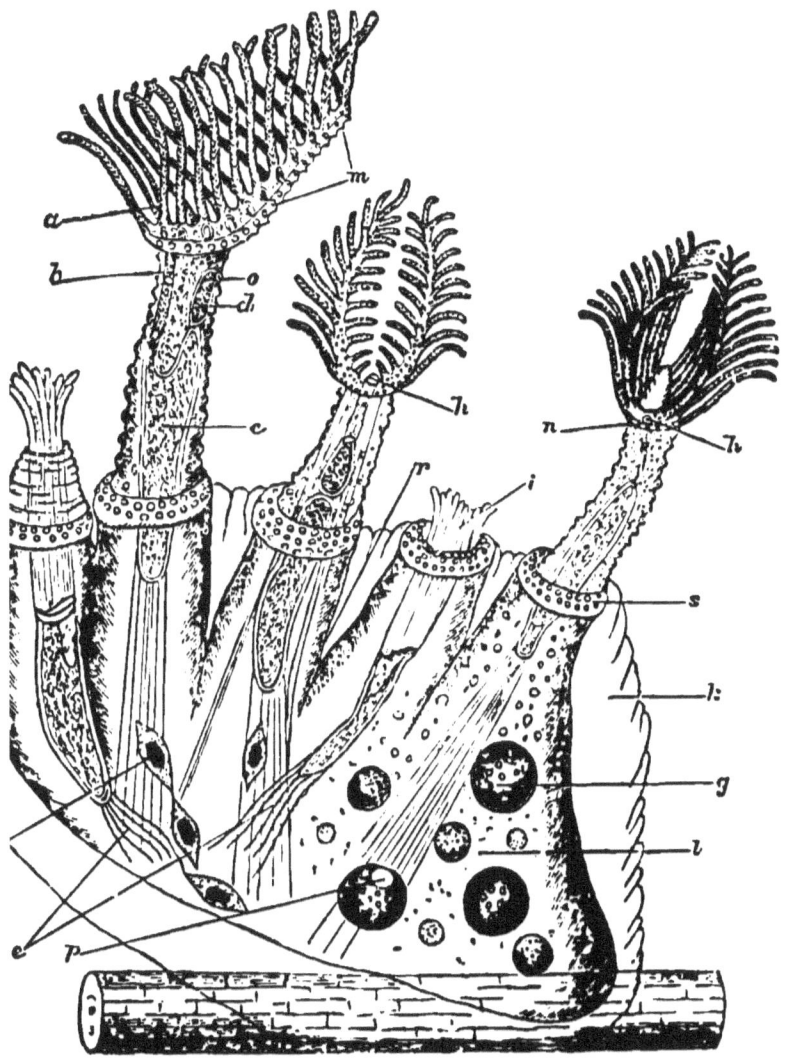

Fig. 281.—Fresh-water polyzoan (*Lophopus crystallina*), magnified.

mass all have in common. The latter is so transparent that you can behold the entire physiological structure and economical arrangement of the colony. When half a score of these creatures have put out their crescent-shaped "lophophores," the sight is like that of some fairy flower-garden.

Fig. 282.—The tunic of dead polyp filled with statoblasts, or winter eggs.

These fresh-water Polyzoa have three methods of reproduction—by eggs, buds, and *statoblasts*. The latter are "*winter* eggs," and they are usually secreted within the body of the polypes. The polypes die, and for a time act as a shelter for these winter eggs. The heat of the returning spring decomposes the body, sets the winter eggs free, and by-and-by hatches them. Each polyzoan colony, therefore, begins with a single individual hatched from the egg.

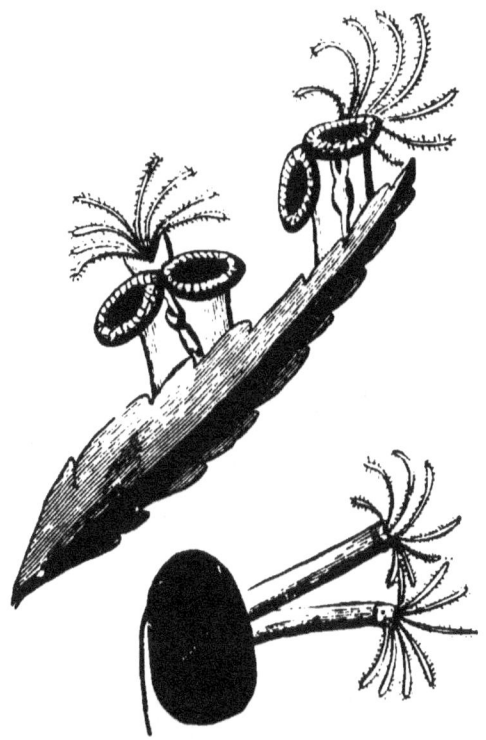

Fig. 283.—Statoblasts, or winter eggs, of Plumatella developing.

How my young friends worked during those summer holidays! Only it was not *called* work. It is calling any occupation by that name, and

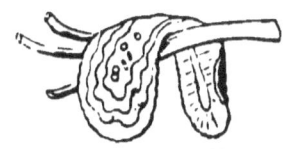

Fig. 284.—*Cristatella mucedo* (natural size).

making people *work* at it a fixed number of hours a day, which disgusts them. Call any real hard

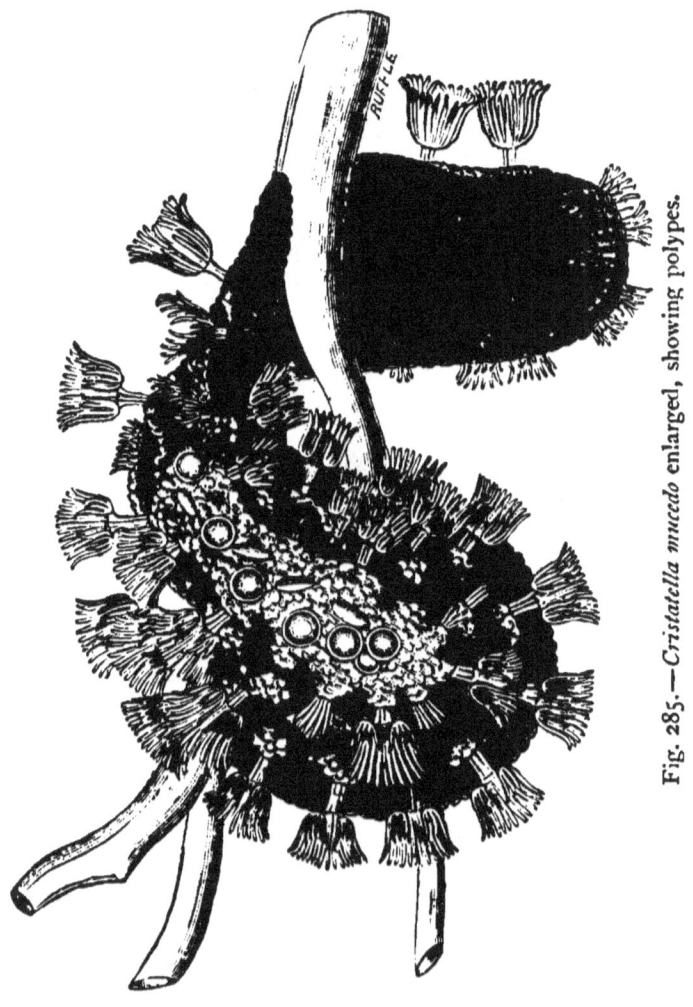

Fig. 285.—*Cristatella mucedo* enlarged, showing polypes.

work *sport*, or *recreation*, or anything of that sort,

and all of us are ready to stick to it till we nearly drop—that is, if we like it.

So, although I am trying to lump each lot of their "finds" separately, just for order's sake, and to act generally as a recording scribe and scientific secretary for my young friends, I find it difficult not to get things a little mixed. For, when the lads were getting Polyzoa, they were also netting

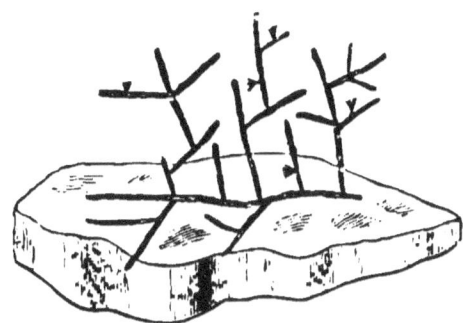

Fig. 286.—*Paludicella sultana* (natural size).

water-worms, larvæ of aquatic insects, and vast numbers of those singular creatures known as the " wheel animalcules," or Rotifers. The latter were everywhere, voracious and active, roaming about like microscopic lions seeking what they could devour. All of them are nearly quite transparent, but some just faintly tinted with pink, or opal blue, or white. As busy as bees, or rather *ants*, everywhere. Most of them swam so rapidly that they came and went before you got the note of admiration out of your mouth! Then an individual would

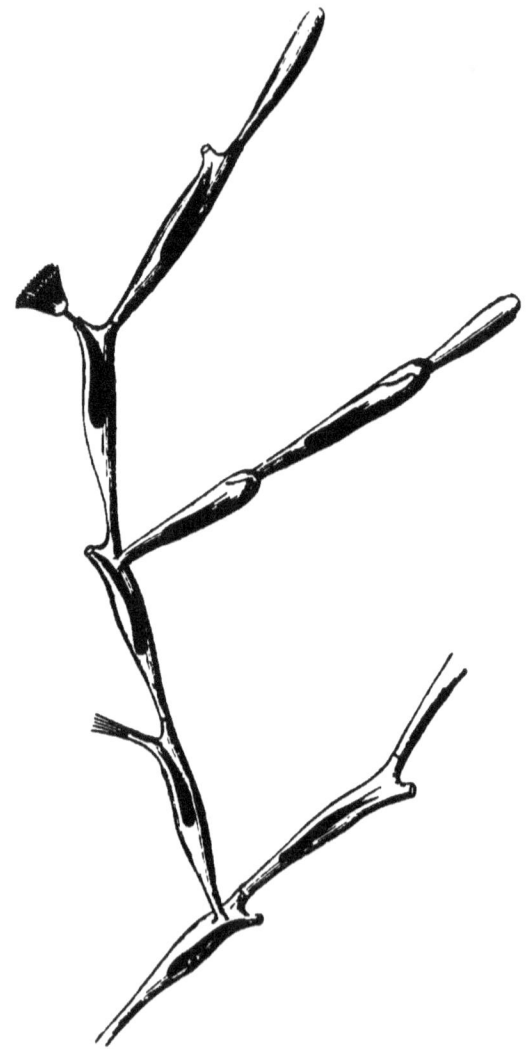

Fig. 287.—*Paludicella sultana* enlarged, showing polypes.

SMALL FRY. 223

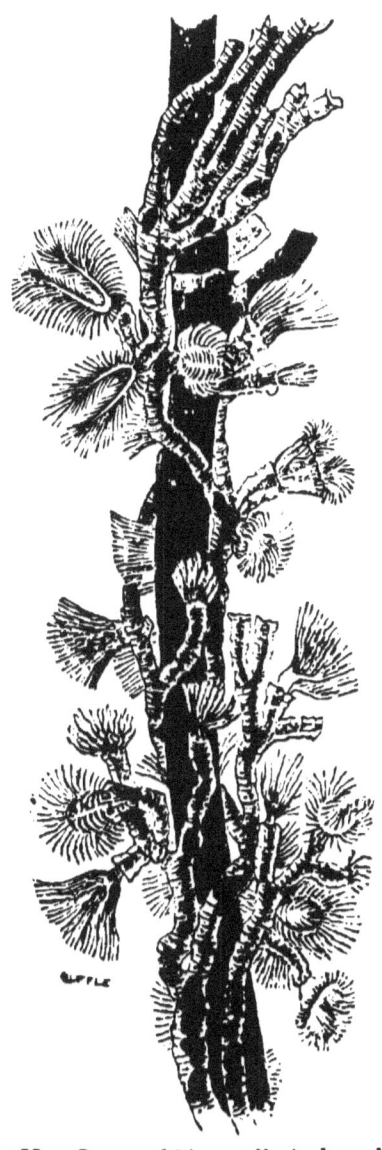

Fig. 288.—Group of Plumatella (enlarged)

boldly crawl or slowly swim right up to the inner surface of the glass trough, as if it were looking through one end of the microscope at you, whilst you were gazing through the other end at it.

Fig. 289.—*Melicerta ringens* (magnified).

You could plainly see the "whorls" round the mouth seeming to rotate like so many cogged wheels. This rotation is only apparent, not real. It is caused by each of the hairs or *cilia* bending or moving rapidly in succession. Why, our own eyelids would appear to rotate if each of the eyelashes of the opened eyelids, upper and lower, behaved in the same way.

But many kinds of rotifers were observed to be

Fig. 290.—Stephanoceros just emerged.

Fig. 291.—First formation of tube.

stationary. They were fixed to weeds, or bits of fine withered stems, or something of that kind. Not a few of them had a sort of sheath into which they could retreat when danger threatened; and

Fig. 292.—Resting period.

Fig. 293.—Appearance later on.

at least one kind, and that one of the very loveliest, made a tube of round pellets out of its own rejected food-materials! I allude to *Melicerta ringens*.

Q

Stephanoceros eichornii is one of the commonest of these sessile and tubed wheel-animalcules. Recently its history has been written and published in *Science-Gossip* by Mr. W. H. Harris, from the egg stage to that of the tubed or enthroned and

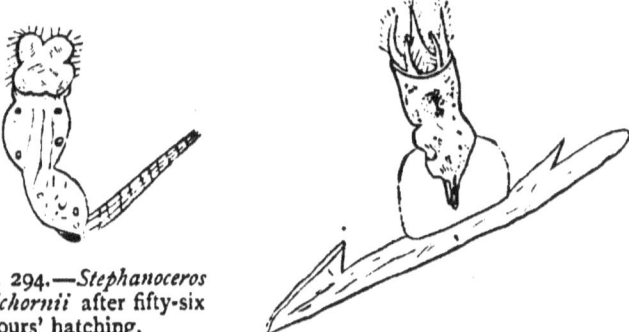

Fig. 294.—*Stephanoceros eichornii* after fifty-six hours' hatching.

Fig. 295.—Later development.

dignified adult. I give illustrations of these several stages of development from the egg. At first, the young Stephanoceros is a free swimmer, from which it appears the sessile state is an advance

Fig. 296.—Suspected male. Scale = 1000 inch
a, Pear-shaped cavity.

on the locomotive or free-swimming state. From the time when a young *Stephanoceros eichornii* animalcule is born, until the time when it acquires the dignity and importance of parenthood itself, ranges from five to nine days, evidently varying

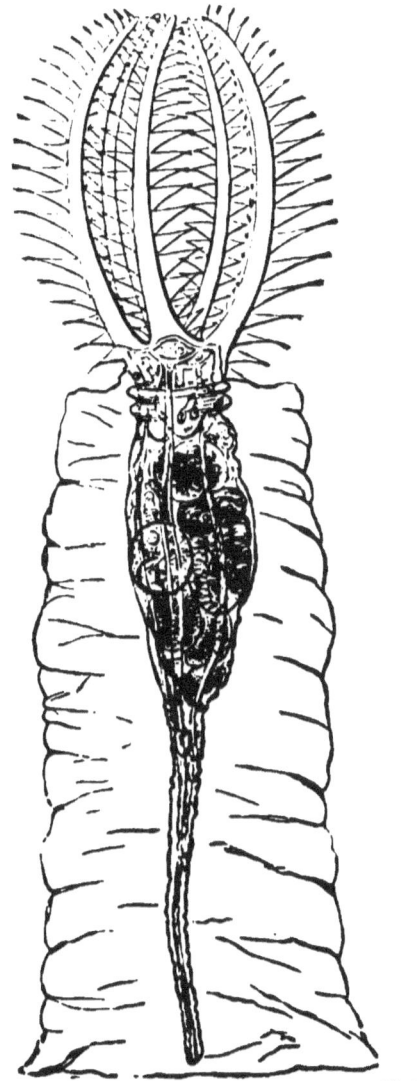

Fig. 297.—*Stephanoceros eichornii* (magnified)

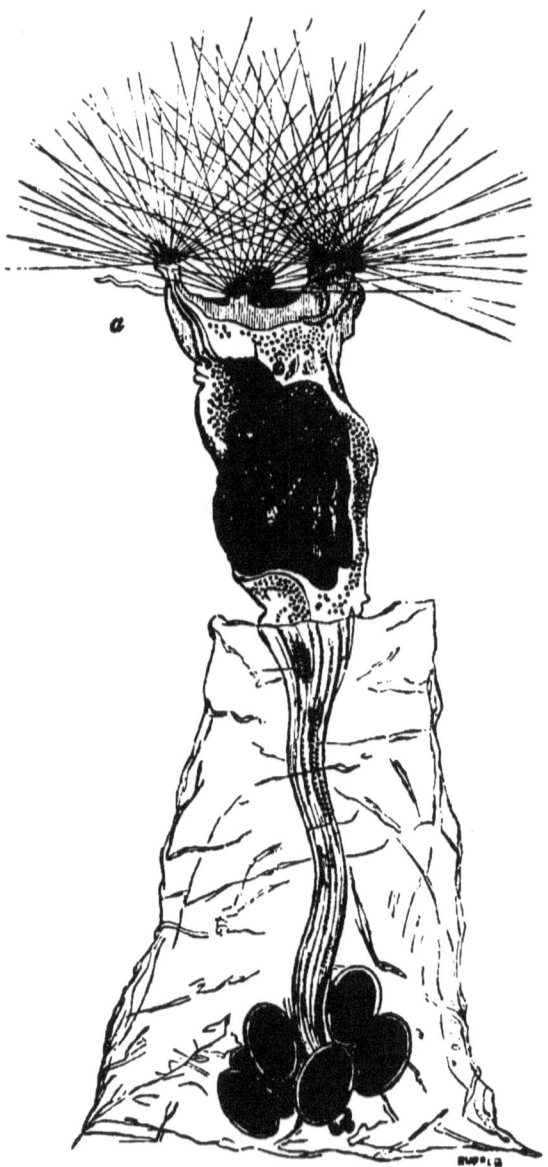

Fig. 298.—*Floscularia cornuta* (magnified).

with the heat—an average of six and one-third days.

You will see by the picture that possibly only the female rotifers are sessile. The males are free; perhaps they prefer it! But, at any rate,

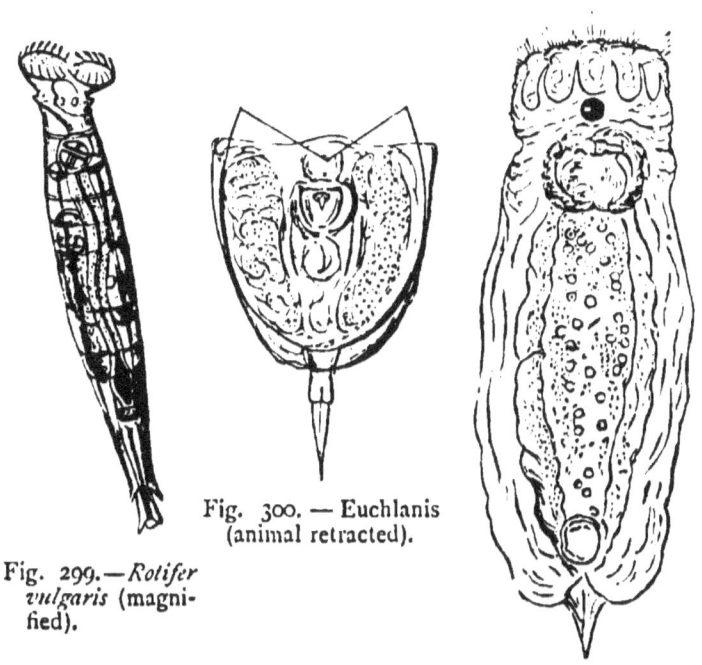

Fig. 299.—*Rotifer vulgaris* (magnified).

Fig. 300. — Euchlanis (animal retracted).

Fig. 301.—*Anuræa leptodon* (magnified).

their wives can upbraid them when they come home with the fact that they belong to a lower stage of rotifer-life, and not a higher. Perhaps this pleases both sexes. The males remain content

230   THE PLAYTIME NATURALIST.

to wander fancy free, and the females to enjoy their superior position. Who knows?

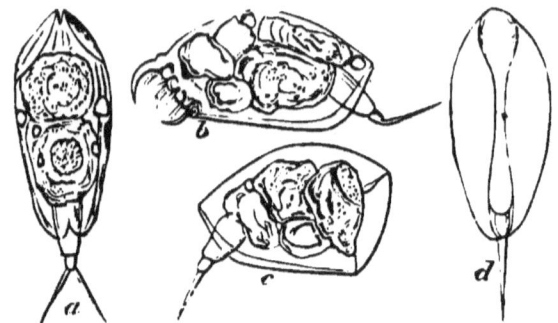

Fig. 302.—*Colurus uncinatus* (magnified). *a*, Dorsal view; *b*, side view, animal swimming; *c*, side view, cilia retracted; *d*, ventral view of lorica.

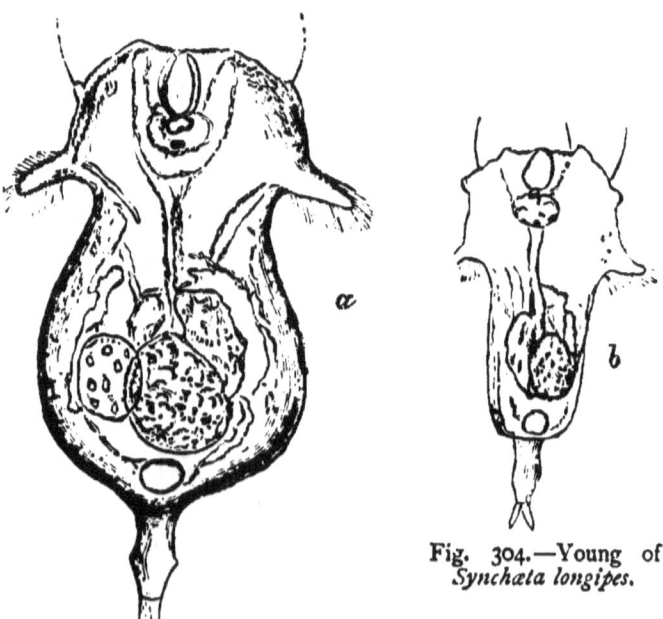

Fig. 304.—Young of *Synchæta longipes*.

Fig. 303.—*Synchæta longipes* (magnified).

SMALL FRY. 231

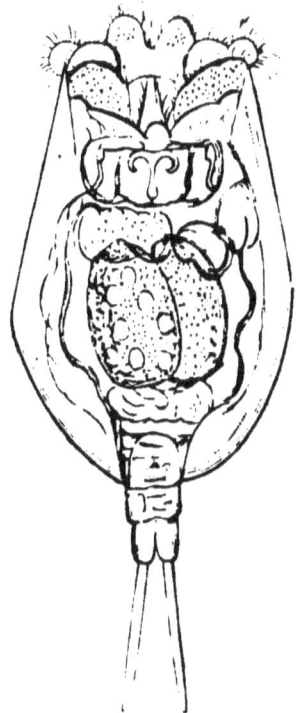

Fig. 305.—Euchlanis (exserted).

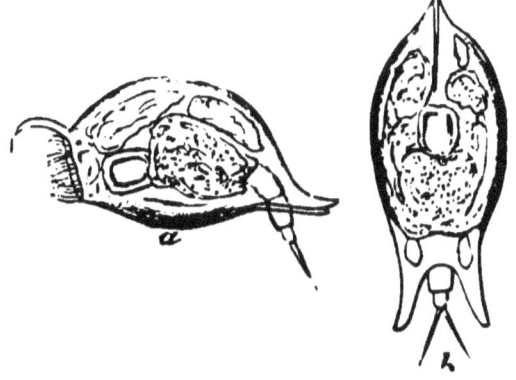

Fig. 306.—*Colorus deflexus.* a, Side view; b, dorsal view.

Among the most queenly looking of these sessile rotifers is the crown animalcule (*Stephanoceros*),

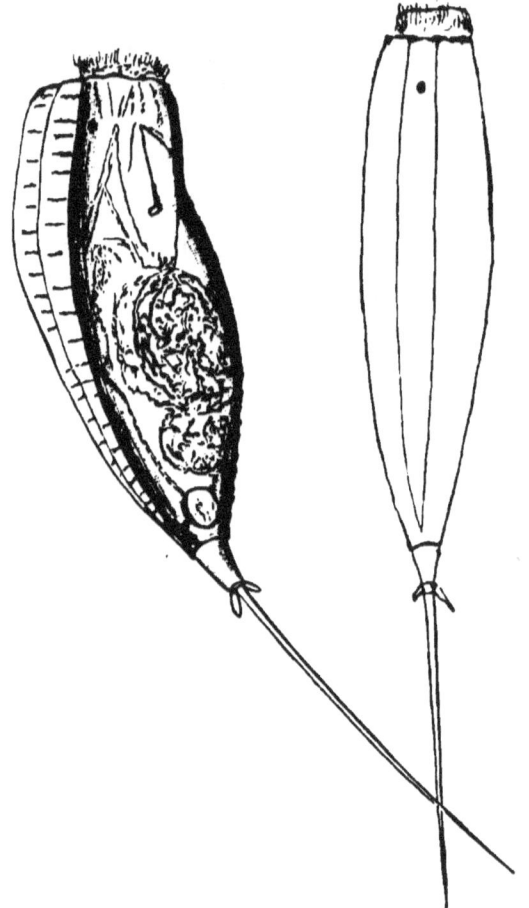

Fig. 307.—*Mastigocerca bicristata* (magnified).

with its transparent glass-like sheath, through which you can clearly trace her majesty even when she has retired from public view; the "Floscule" (*Floscularia cornuta*); and others. Among the commonest of the "free rovers" and never-settlers-down, here, there, everywhere, making every zoophyte-trough as lively as a ballroom with their goings-on, are the common rotifer (*Rotifer vulgaris*), Synchæta, Euchlanis, Colorus, Mastigocerca, and others. In most instances it was observed that the males and females possessed distinctive characters, and that the females seemed to outnumber the males. Indeed, in the case of many common female rotifers, we don't know who their husbands are, or what they are like.

CHAPTER XI.

INVISIBLE LIFE.

"I SAY," said Jack Hampson to his friends, "suppose those wild rotifer beggars were really as big as they looked in the microscope, wouldn't people be frightened?"

"You should see them living in a thin zoophyte-trough and magnified, and their images thrown on a screen, like the pictures in a magic lantern! You've no idea how horrid they look. I've heard elderly gentlemen say they would never drink water any more, when they saw them," said Willie.

"Perhaps they didn't drink much before," remarked his friend.

But there remained over for observation a vast, unnamed, and unrecognized army of living beings, fellow-creatures, sitters-down at the same wonderful providential table as ourselves, remaining to be recognized and identified; not a few—perhaps most of them—the Lazaruses which devour the crumbs that fall from Dives' tables, and so keep the floors of the houses in which the latter dwell, all the cleaner.

One evening they were examining under the microscope a hydra they had found that morning. They were not thinking of anything but the hydra, and were looking at it in turns, when Jack suddenly said—

"What are these little brutes running all over its body? They look like new telegraph messengers who don't know where to deliver their telegrams."

Willie rushed to the instrument at once. There was a half-inch objective on, rather high for a hydra. After a minute or two's careful examination, Willie said, "Oh, it's an infusorian!"

"What's an 'infusorian'?" said one of the group "You speak as if you knew the whole family. Did they come over with William the Conqueror?"

Willie was not a jokist, so he said, "No; they were here before William the Conqueror's time."

"Then tell us all about them, or, at least, all *you* know about them."

"Well," said the earnest lad and truthful, "that isn't much; but, at any rate, I'll tell you all I've known and heard about them from dad."

So he proceeded to relate how the Infusoria,

were a class of the lowest organized of living creatures, and probably the very oldest in the world, only, having no shells or other solid parts, they were not geologically represented in the oldest strata, although some very nearly allied relatives of theirs were, which happened to be possessed of such things—the Foraminifera, to wit. But even as regards them, the Amœbæ were more nearly related, he said.

"What are the Amœbæ? Are they *girls?* It sounds like a girl's name," remarked one of the newly arrived and irreverent cousins.

"They are girls and boys both," replied the youthful savant, "for they are neither male nor female."

"Why can't you call these things by names that people understand, instead of using such jaw-breaking names as Infusoria? What are they?"

"You ask me two different questions at once, but I think I can put you right. 'Infusoria' is an old word really, given to all sorts of exceedingly minute living objects found in water. It was discovered that when there was an *infusion* of any organic matter, such as decomposing vegetation, the nasty water soon swarmed with myriads of these objects. People used to believe they were bred by the stagnation, and that they were spontaneous or self-created generations. They were only found abundantly where such decaying substances were mixed with water—in *infusions*. So they came to be called *Infusoria*. The decaying materials are their food ; you couldn't keep them in absolutely pure water, for there would be nothing in it for them to live on, and they would die of starvation."

"Then, do you mean to say these things *clear* the water?"

"If it were not for their clearing the water by devouring the amazing quantities of decaying animal and vegetable substances it contained, the world would soon be unfit for larger and much more highly organized creatures to live in it, that's all. These minute living things are the scavengers of the Almighty!"

Then it was explained how, half a century ago, when any minute thing was seen moving or wriggling in water under microscopical examination, it was set down as an *animalcule*, for people then regarded locomotion as an essentially animal function. How, for this reason, in such books as Pritchard's "History of British Infusoria," the zoospores of sea-weeds, and similar structures connected with mosses, ferns, etc., were put down as "infusorial animalculæ," because they could swim and move about. Of course, this is all set right now.

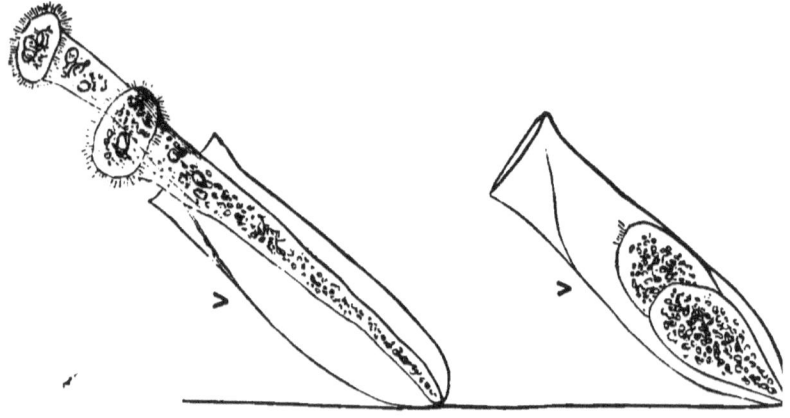

Fig. 308.—Vaginicola extended and withdrawn (magnified).

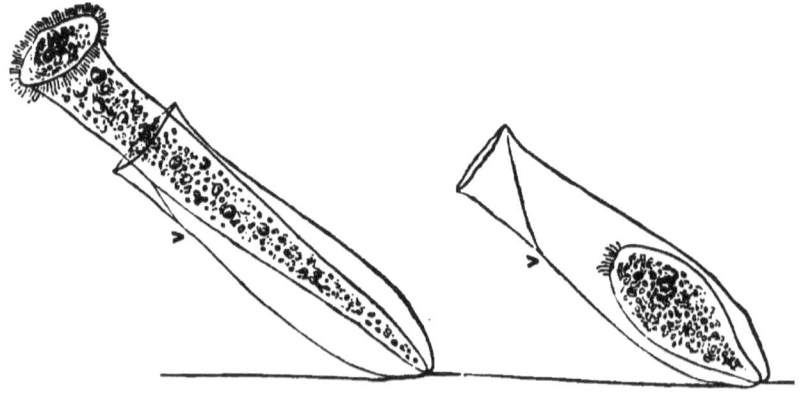

Fig. 309.—Vaginicola after fission (magnified).

But the remarkable fact remains that very probably every animal organism, not excepting man, begins life in an infusorial stage. That is the "narrow wicket-gate."

The hydras were the subject of amusing observation in the evenings on account of the peculiar movements of their snake-like tentacles, and the rapid skating over the surfaces of their bodies

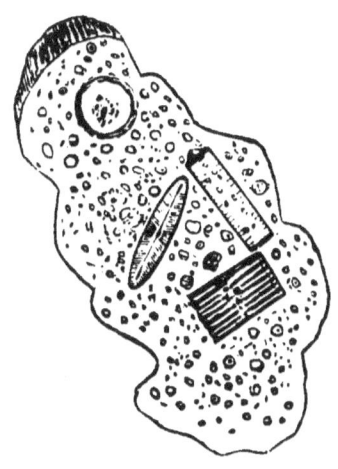

Fig. 310.—*Amœba villosa*.

Fig. 311.—Infusorial parasite of hydra.

by the aforesaid animalcules, or "parasites." It was seen that the organs of locomotion possessed by the latter were eyelash-like hairs (*cilia*), or prolongations of the transparent body-substance drawn out into short short sensitive threads.

"Well," remarked one of the sceptics, "I've heard of a fellow 'hanging on by his eyelids,' but never of his *walking* on them before, much less *swimming* with them. But, according to your tale, that is what they do with their cilia, and *cilia* means 'eyelashes.'"

Somebody remarked that it was just possible there were several other things the gentle youth

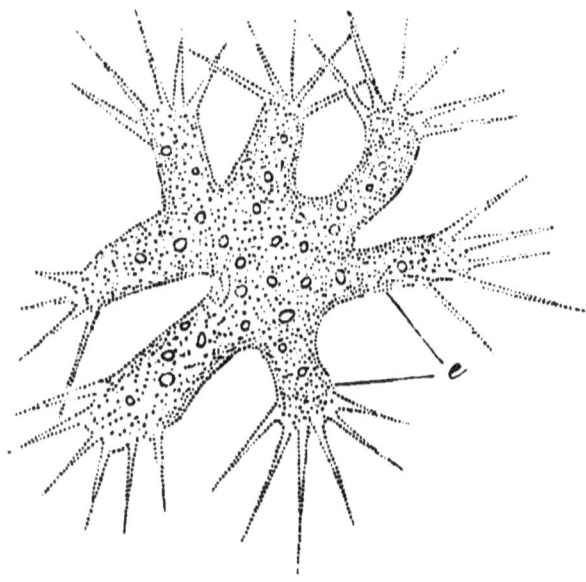

Fig. 312.—Amœba with compound pseudopodia.

had never heard of before, which he probably would hear of before he was as old again—that is, if he kept his ears open.

The amœba seemed to be the chief favourite. Willie repeated the speculations he had heard when his father had one or two savans to supper—

how probable it was that the amœba really represents the first animals that came into this world.

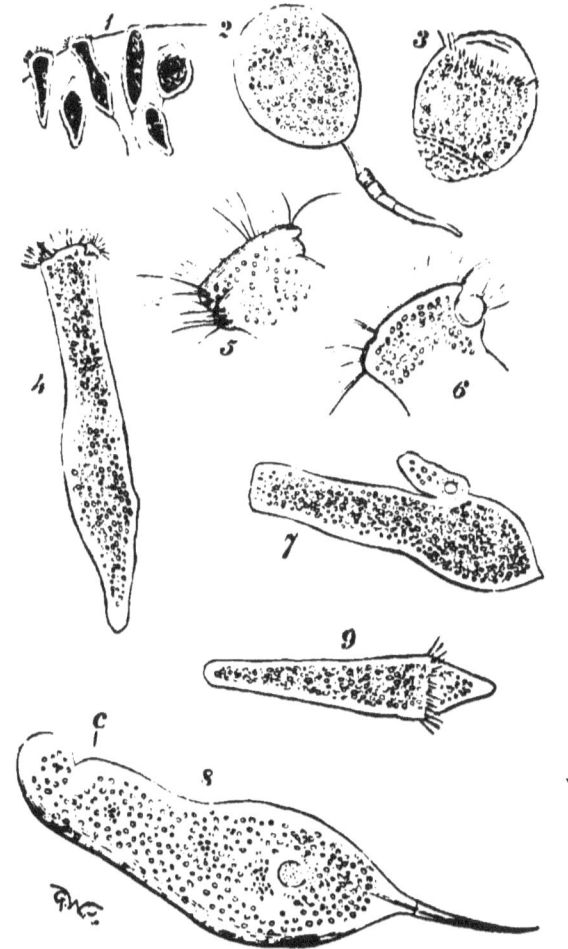

Fig. 313.—Jelly animalcules (*Ophyridium versatile*). 1, Group in gelatinous envelope; 2, 3, 4, separate individuals in different conditions; 5, 6, head magnified; 7, young animalcule produced by gemmation; 9, swimming animalcule.

How the Foraminifera, which had formed limestone and chalk strata thousands of feet in thickness, were in reality only amœbæ with *limey* skins; how the white corpuscles in the blood of man and other animals could hardly be distinguished from them; and how many of the fungi (Myxomycetes) began life practically as amœbæ;—indeed, how this lowly organized stage of structure was so common in the animal and vegetable kingdoms, that the word "amœboid" had been coined and was in constant use, to express the fact.

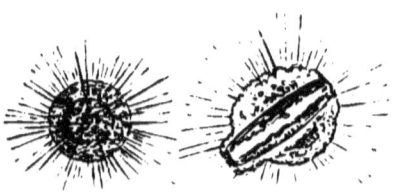

Fig. 314.—*Actinophrys sol.*

These amœbæ were everywhere, said the books they referred to. But it was a long time before they found one; afterwards they found any quantity. Jack and Will were dreadfully anxious to discover these abounding amœbæ, and could not find one—which, of course, made them all the more eager in hunting.

Still, they failed to find an amœba. The fact was, they didn't know what to look for. So they wrote to their old (or rather young) professor, as they always did when they came to a sudden stop.

He replied that the amœbæ looked more like minute irregular splashes of transparent water, spilled on the outside of the glass trough, than anything else. They were merely specks of transparent, living jelly—almost exactly like the raw

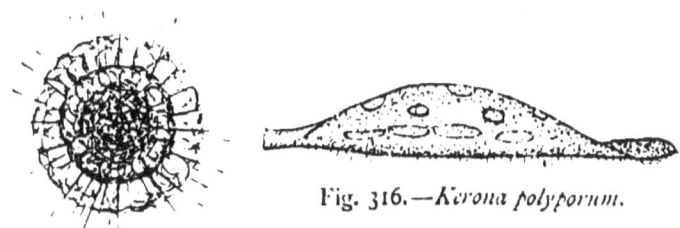

Fig. 315.—*Actinophrys aculeata.*

Fig. 316.—*Kerona polyporum.*

white of egg; only they could move about as they liked, and a speck of white of egg could not. The best plan for catching amœbæ, he said, was to lower a zoophyte-trough down into the aquarium at night. Then haul it up next morning; they would find plenty of amœbæ in it.

The boys did so, and were delighted in "spotting" their first amœbæ. After that they found them everywhere—"all over the shop," Jack said. They watched them slowly moving towards a decomposing bit of animal or vegetable matter; then pouring themselves over it as if they were merely animated gum, until the dainty morsel was *inside* the amœba—which is where food ought to be. So, if these singular animals have neither mouths nor stomachs to introduce their food into, they manage

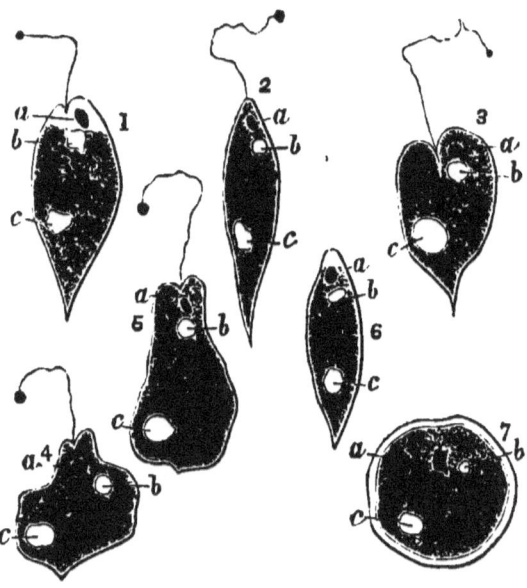

Fig. 317.—Various stages in the development of *Euglena viridis*.

to put themselves outside their food—and I believe that is the chief end of animal life.

Then there were the Infusoria to be caught and watched. They soon learned that the discoloration of any pond-water, or even of lake and sea-water was due to them. They cause the summer phosphorescence of the sea, and the remarkable green, brown, and reddish tints seen in most natural waters at times. They found the water of one pool quite brown, and a microscopical examination proved that it was entirely due to swarms of a special kind of infusorian known as Peridinium. Another pond possessed a vivid green-coloured water, and this they soon proved was owing to the countless numbers of *Euglena viridis* in it— a pretty green infusorian, with a brilliant red "eye-spot" (Fig. 317).

Further, they speedily discovered that the Infusoria, like the Rotifera, could be separated into "free-swimming" and "sessile," and that the earliest life-stages of the latter resembled the former.

They found out the exquisitely shaped "Greek vases" of the Vaginicola, more transparent than any glass, into which the dainty microscopical marvels withdrew themselves at will. They discovered the still more beautiful clusters and colonies of Vorticellæ, like bunches of lilies, and watched their sensitive stalks twist and untwist like living corkscrews. They knew now what those strange tufts were on the heads of the water-fleas (Daphnia), and that they were stalked infusorians allied to the Vorticellæ, called Epistylis.

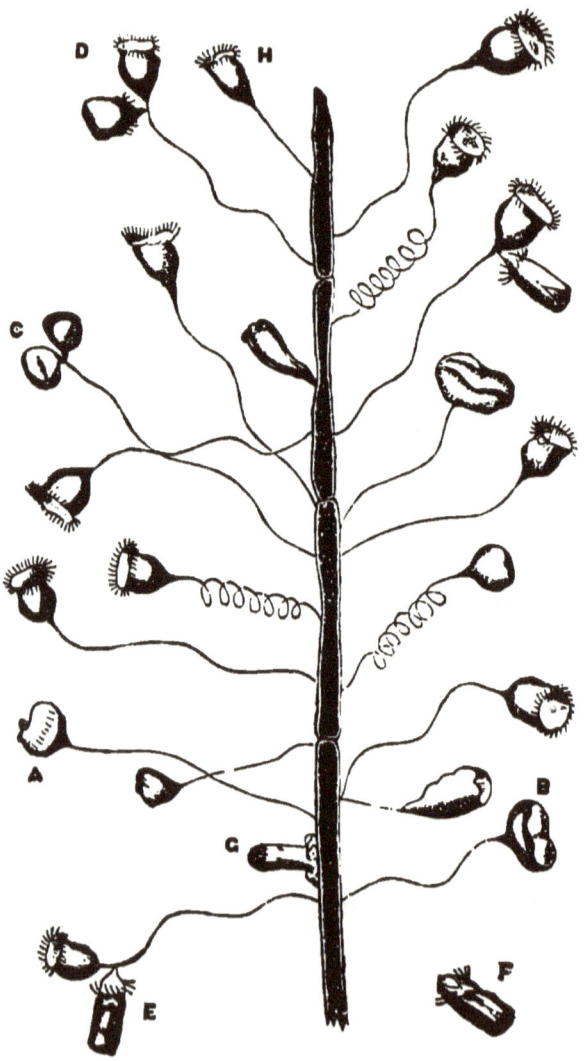

Fig. 318.—*Vorticella nebulifera*, showing development of individual stages A to F (E and F free).

They even made out the different stages of development of many of these lowly organized objects

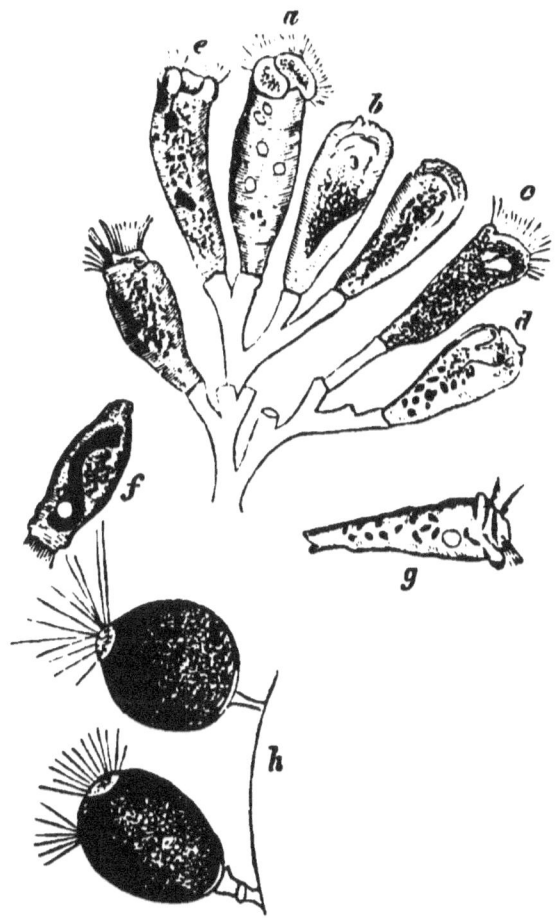

Fig. 319.—Various stages in development of Epistylis, semi-parasitic on water-fleas.

—their larval, resting, and adult conditions—in many cases differing so much from one another that they have been figured and described as distinct species, and as altogether different creatures. What a host of things of this kind there are! Sluggish Actinophrys, or "sun-animalcules" (difficult to be distinguished from the resting stages of other infusoria), clusters and crowds of

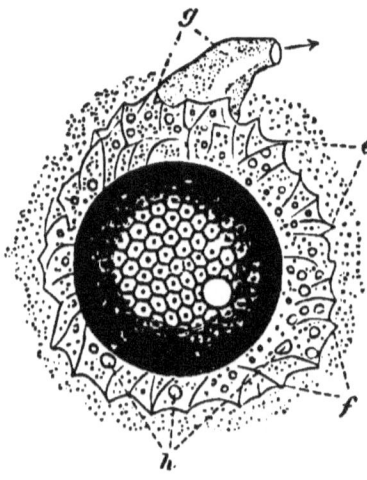

Fig. 320.—Ovarium of fresh-water sponge. *e*, Growth of spiculæ; *f*, sarcode festoons on ditto; *g*, mouth; *h*, incurrent pores.

certain kinds which seem to be born, or rather to be reared, together—microscopical "baby-farming;" fresh-water sponges (*Spongilla fluviatilis*), dredged up from the bed of the river, clinging to and covering up dead twigs with their greenish gelatinous matter. The ovaries, or egg-bearing chambers, of the latter were found; also the remarkably beautiful spicules which encase them. The young of these fresh-water sponges were

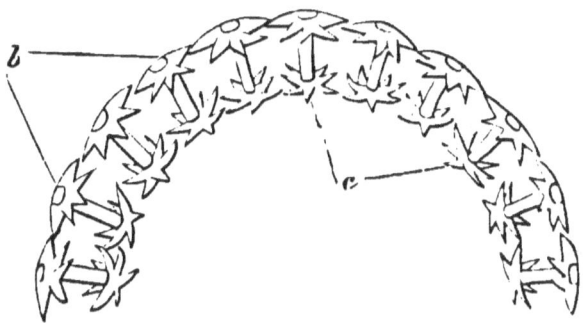

Fig. 321.—Rotate or wheel-shaped spiculæ. *b* support the outer membrane ; *c*, inner ditto.

obtained and reared, and a good deal of the life-history of this interesting and instructive animal, or rather colony of animals, was made out in the course of observation and investigation.

Don't suppose for one moment that all these discoveries were made by my young friends without external help. On the contrary, they had plenty of assistance ; as every young naturalist will find

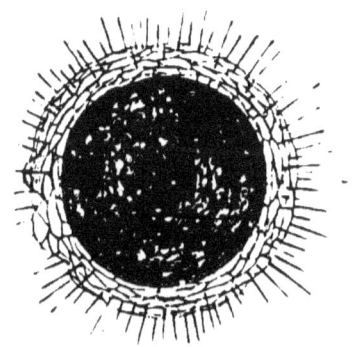

Fig. 322.—*Actinophrys eichornii.*

he can get, if he is only in earnest. For there is a brotherhood among naturalists like that among artists, but without anything of its jealousy. The boys got letters from Willie's dad, from the professor, from the editor of *Science-Gossip* (to whom they consigned loads of specimens), and others. It was awfully jolly while it lasted.

CHAPTER XII.

MICROSCOPIC PLANTS.

THE longest of summer holidays comes to an end. Ours were now over, and the boys had returned to school. Willie left a week before Jack, to spend the rest of his time at home. You may depend upon it, there was much comparison of captures and notes, and much examination of specimens. The science-classes at the school had never been so popular before.

The Natural History Society resumed its meetings, and fresh papers were read, some of them dealing with the captures of the holiday-time. Among others was an impprtant paper on those singular and beautiful microscopical plants, the Desmids and Diatoms. As the autumn was not too far advanced, the afternoon holiday rambles were in force, and plenty of new "finds" made. They could hardly wander a yard into the country without finding something they had never seen or known before—perhaps the dense brownish-black fungus covering the under surfaces of the thistle-

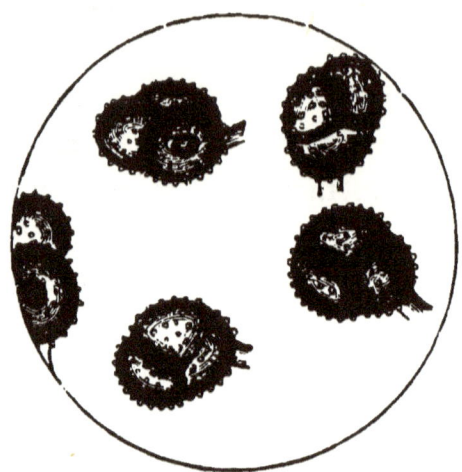

Fig. 323.—Meadow-sweet brand.

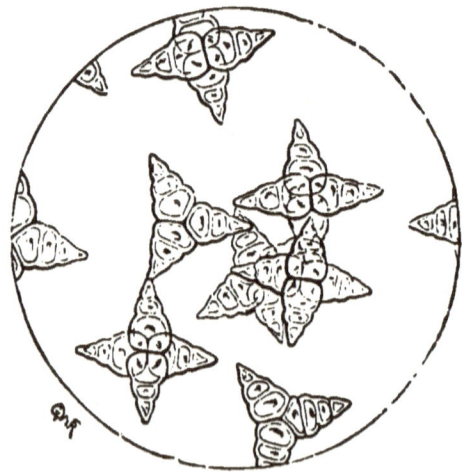

Fig. 324.—Star-spored brand.

leaves in the pastures; or that other equally common but different kind which attacks the leaves of the mallow, and drills them full of round holes. There were also the pretty fungus, or "brand," of the meadow-sweet; the star-shaped fungus found on dead twigs (Asterosporium); the

Fig. 325.—Bramble-leaf brand.

exquisitely pretty bramble-leaf brand; the maple-blight, etc. Among other objects which could not fail to attract their attention were the numerous glossy black spots on the leaves of the sycamore. These were formerly believed to have been caused by drops of water acting as lenses for the sun's

254   THE PLAYTIME NATURALIST.

heat, and, by focussing the heat, burning and blistering the leaves. But the professor showed that this fanciful notion was altogether wrong, and that the black spots were due to a peculiar fungus called Melasmia, whose structure and character

Fig. 326.—Maple blight.

could be made out by cutting a section of sycamore-leaf across one of the black spots.

Even more funny as an explanation than the above is the assumed origin of the gelatinous masses found on the gravel walks in our gardens on September mornings (*Nostoc commune*). They

MICOSCOPIC PLANTS. 255

Fig. 327.—Sycamore-leaf with black spots of *Melasmia acerina.*

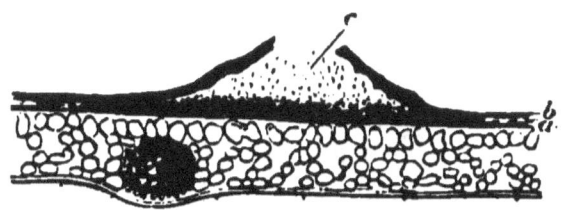

Fig. 328.—Section through leaf, showing position of the fungus, *c.*

have actually been set down as "shooting-stars" which had fallen to the earth during the night! A microscopical examination however, shows their

Fig. 329.—Witches' butter (*Nostoc commune*).

neck-beadlike arrangement of cells, and establishes their fungoid structure.

Some of the boys had taken up with the larger

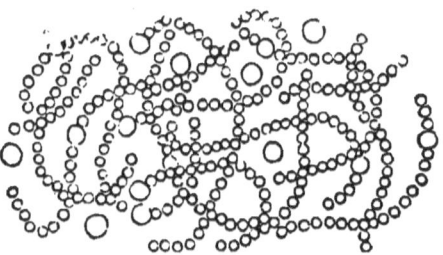

Fig. 330.—Spores and cells of ditto (magnified).

funguses, and were collecting them and getting them named. They got Mr. English's book, which told them how to preserve these objects—hitherto so exceedingly difficult to preserve—so as to keep their shapes and colours, and altogether form pretty ornaments, when under glass shades, for rooms. A large number of specimens were got, which I cannot stay to describe; nor is it necessary, for the professor had in his study those two mounted

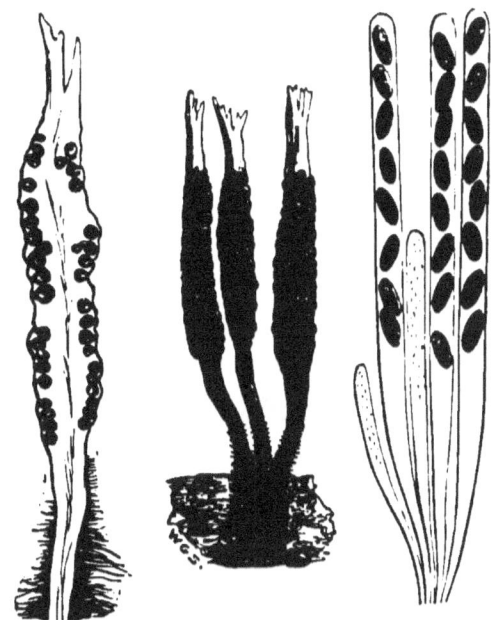

Fig. 331.—Candle-snuff fungus (*Xylaria hypoxylon*).

and coloured sheets of "Edible" and "Poisonous" fungi by Mr. Worthington G. Smith, which show all common kinds at a glance. One kind, however, interested them much—the "candle-snuff" fungus (*Xylaria hypoxylon*). Plenty of it was found in the damp hedge-backings, and its microscopical structure afforded a pleasant evening's intellectual pastime.

But the main objects of the rambles in September were to collect desmids before the season had advanced too far. It was determined to collect the diatoms as well, for both occur in the same tarns and ponds, only you can collect the latter at any time of the year, and this is hardly possible with the desmids in the winter months.

As the party walked on their way, the professor explained the structure of the collecting-bottle he carried with him. It was merely a wide-mouthed, one-ounce bottle, provided with a turn-back rim. Around the latter was a strong india-rubber band. He showed them he had only to double the elastic round the end of his walking-stick, and he was able to push it anywhere along the margin of the pond. He had several others, all provided with close-fitting corks, for specially keeping good things in until he could further examine them. For collecting diatoms he was provided with a

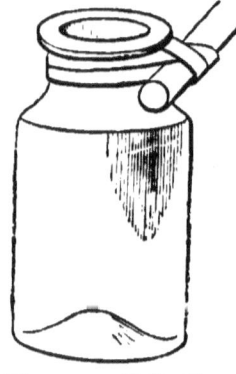

Fig. 332.—Collecting-bottle.

more elaborate apparatus, but the cleanliness and freedom from impurities with which diatoms could be collected by it, he said, made it the best thing of its kind out; and everybody could make one now that Mr. Redmayne had shown them how.

This diatom collecting-bottle is constructed as follows:—

A cork must be provided which fits tightly to the collecting-bottle; this is to be bored with two holes. In each is fitted a glass tube, as seen in the diagram, one (a) having a slight curve, the other (b) bent at right angles an inch from the end; this can easily be done with the aid of a spirit-lamp. To tube b is attached a piece of elastic tubing, about the length of the collecting-stick, and the free end (c) may be held to the stick with an elastic band, and the apparatus is complete.

It is especially useful in collecting the very thin films of diatoms from the surfaces of mud and sand, so difficult to raise to the surface of the water in the ordinary way with the spoon or bottle.

To use the apparatus, the thumb of the right hand must press the tube firmly against the stick at c, and the bottle be lowered until the mouth of the tube (a) is within a quarter of an inch from the surface of the diatoms; the thumb is then raised, and if the water is deep the bottle fills by atmospheric pressure, carrying the diatoms in at the same time. In shallow water, suction will be necessary to exhaust the air in the bottle; in

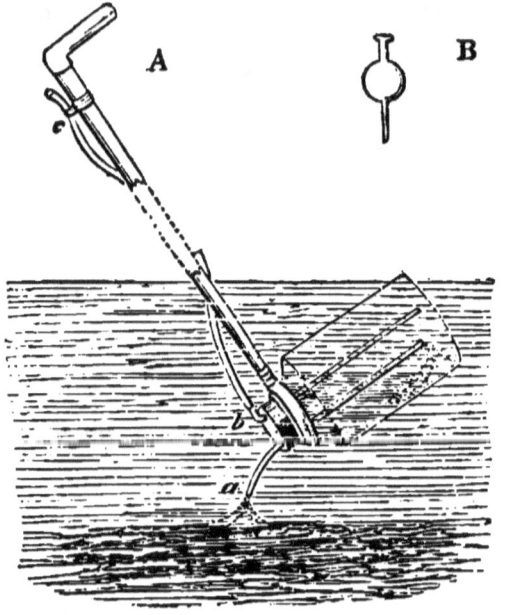

Fig. 333.—Diatom collecting-bottle.

that case, a ball pipette (B) will be useful as a mouthpiece.

The gathering can be further cleaned by placing it in a glass bottle in the sun for a few hours. Cover the lower part of the bottle with black paper; the free diatoms will then separate themselves from the mud, and rest on the surface.

MICROSCOPIC PLANTS. 261

To collect desmids, it would be best perhaps to use the first-mentioned collecting-bottle. Our boys easily learned to detect the appearance of the vivid green desmids from the olive-brown appearance of the diatoms. Further, the desmids love the surface of the water where the sun can get to them; whilst most of the diatoms prefer the shady bottoms.

Desmids were quickly found—the common species in abundance. What pretty little plants they are, as green as spring grass, and possessing a transparent greenness you will not find in any other kind of vegetation! They prefer clean, sweet water. A mountain tarn is a place they love best. I have seen forty distinct species collected from one such spot in North Wales, and you may guess the smallness of these plants when I tell you that all were mounted within the ordinary half-inch circle of a slide. My "show" pictures indicate the magnifications of each kind, and will also give an idea of their minuteness.

Each desmid possesses a transparent case, usually free; but also not unfrequently attached. This case contains the green colouring matter, exactly as the cell of a green leaf does. Desmids, therefore, represent single free cells. Their mode of propagation is by *fission*—that is, each desmid splits into two, the two originate four, four eight and so on. If these "cells" cohered together, the result would be a leaf or some other vegetable structure; but as each is "free," no coherence of that kind takes place, and therefore no "growth"— or increase in bulk.

You can hardly pick the wrong place for desmids, in spite of their preference for clean water—ponds, ditches, rivulets, even the miniature tarns made by footprints of cattle in marshy places; anywhere except *salt* water (for they are purely fresh-water plants, whereas diatoms live in the sea, and in brackish water, as well as under the same fresh-water conditions as the desmids).

One might almost declare the desmids are the food-stock of all fresh-water animalculæ. They furnish an abundant foraging and hunting ground to myriads of infusoria, rotiferæ, aquatic worms, larvæ, etc.

Among the commonest, but not the least beautiful forms are Closterium, Euastrum, Cosmarium, and Micrasterias. There are about forty British species of Closterium alone. If the end of the frond be highly magnified, the green granules are seen circulating at the end. In this genus there is a double and perhaps a treble method of reproduction—by self-division (fission), conjugation, etc.

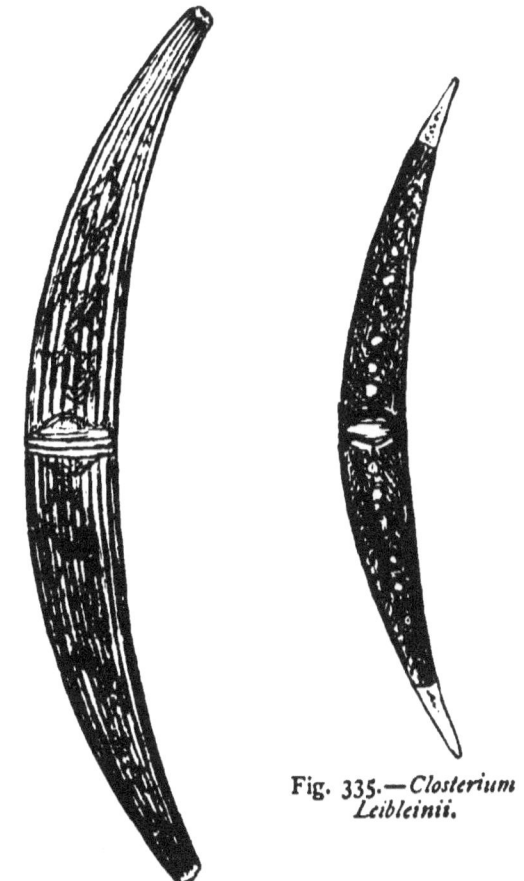

Fig. 335.—*Closterium Leibleinii.*

Fig. 334.—*Closterium striolatum* (magnified).

In every frond of Closterium you observe a central clear space dividing it into two segments.

Here a gradual separation takes place, occupying some hours before it is completed. The separated halves then each commence to grow independently, till ultimately a copy of the parent form is assumed. This is an outline of self-division. Conjugation is a different process. Two individuals approach each other and come into contact. They intermingle their green contents, and a curious globular body is formed, called a sporangium, which is believed in due time to produce a multitude of individual spores, which ultimately grow into Closteria. The operation of forming a sporangium is said to be very rapid, only occupying a few minutes.

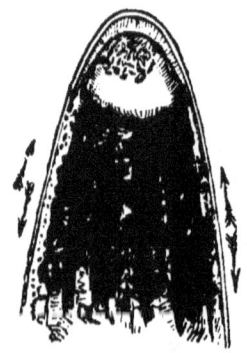

Fig. 336.—End of frond of *Closterium lunula* (highly magnified).

A writer in *Science-Gossip* for 1866 says, "The other evening I saw the end of a bright green Closterium seized by a large animalcule, *Notommata myrmeleo*, and subjected to the action of the teeth. Soon I found that the particles of chlorophyll were leaving the desmid and passing down the gullet of the animalcule, evidently by suction, and I watched them with great interest—first, because I never before saw a rotifer taking a salad in so civilized a manner (they generally take their vegetable diet into their crops by a rapid jerk, particularly when it is small enough to go down whole); and, secondly, because apertures at the ends of the fronds are not generally believed in. When the animalcule had finished its supper—that is to say, when every part of nutriment was gone—it cast the empty frond among others that were strewed about, and I could not detect the slightest rupture in the delicate transparent case, which a few minutes before was so full of green contents. There may have been one, nevertheless."

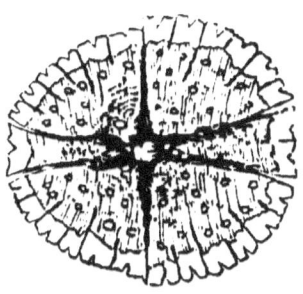

Fig. 337.—*Micrasterias rotata.*

"It is astonishing how long you can keep these desmids," said the professor. "I've kept them for six months even in closed bottles, in the sunlight, or daylight at least; and it is equally astonishing how rapidly they increase."

One of the "sweetest" of these microscopical gems is *Micrasterias rotata*—a flat, almost oval-shaped object, with a delicate transparent frill surrounding the disk of bright green.

Some of the desmids affect a social life, such

as Hyalotheca—rightly named so, for the packed green desmids look as if they were enveloped in

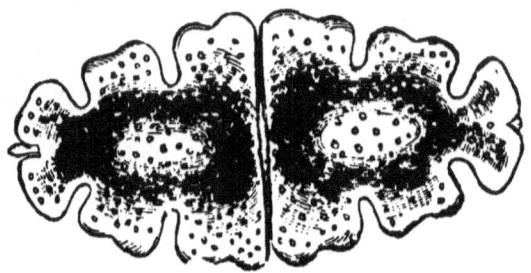

Fig. 338.—*Euastrum oblongum* (front view, × 250).

Fig. 339.—*Euastrum oblongum* (side view).

Fig. 340.—*Cosmarium margaritiferum*.

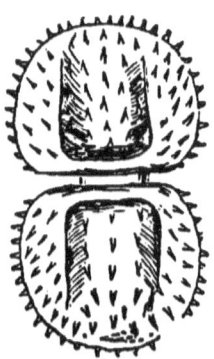

Fig. 341.—Ditto (empty frond, × 250).

a sheath of glass. Then, through such freshwater confervæ as Ulothrix—to be gathered everywhere

—and the exquisitely lovely *Volvox globator* (a perfect vegetable marvel, cell within cell, like Mr. Boy's electrical soap-bubbles), Spirogyra, and other

Fig. 342.—*Euastrum oblongum.*

Fig. 343.—*Euastrum margaritiferum.*

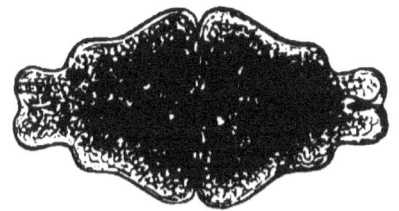

Fig. 344.—*Euastrum didalta.*

  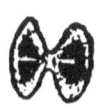

Fig. 345.—*Staurastrum dejectum.* Fig. 346.—*Staurastrum alternans.*

fresh-water algæ, we ascend to those complexer vegetable forms which are complexer because the cells they produce cohere together, cause increase of bulk, division of labour, specialization of function, and all that constitutes higher organization.

In the Ulothrix my young friends soon learned to see that the cells were in reality social desmids, and that some of these were allowed to go forth

Fig. 347.—*Staurastrum spongium.*

Fig. 348.—*Staurastrum gracile.*

free at certain times, and return to the habits of their ancestors, for reasons which were of benefit to the colony—viz. reproduction.

In Spirogyra (almost like Hyalotheca) there is

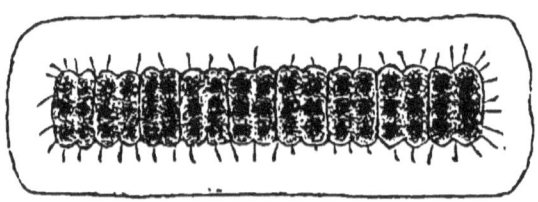

Fig. 349.—*Hyalotheca dissiliens.*

a very pretty arrangement of green chlorophyll, but in bands of spiral filaments, all enclosed in a similar transparent sheath. The cells of some of the bands bud forth, and manage to form a junction, and to interchange their cell-contents. When this is complete, the combined contents of the two cells become an oval spore, from which a new plant

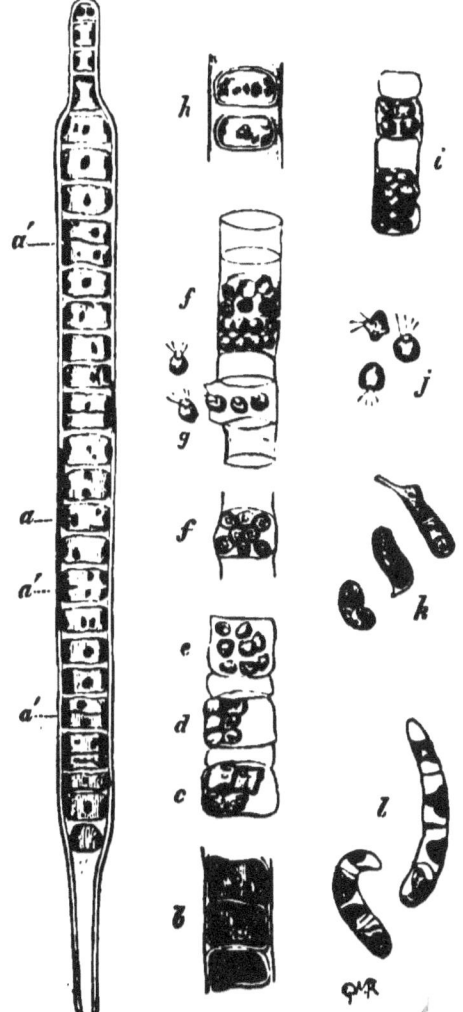

Fig. 350.—Ulothrix. *a*, Young filament; *g*, ciliated zoospores; *k*, one day's growth; *l*, two days' ditto.

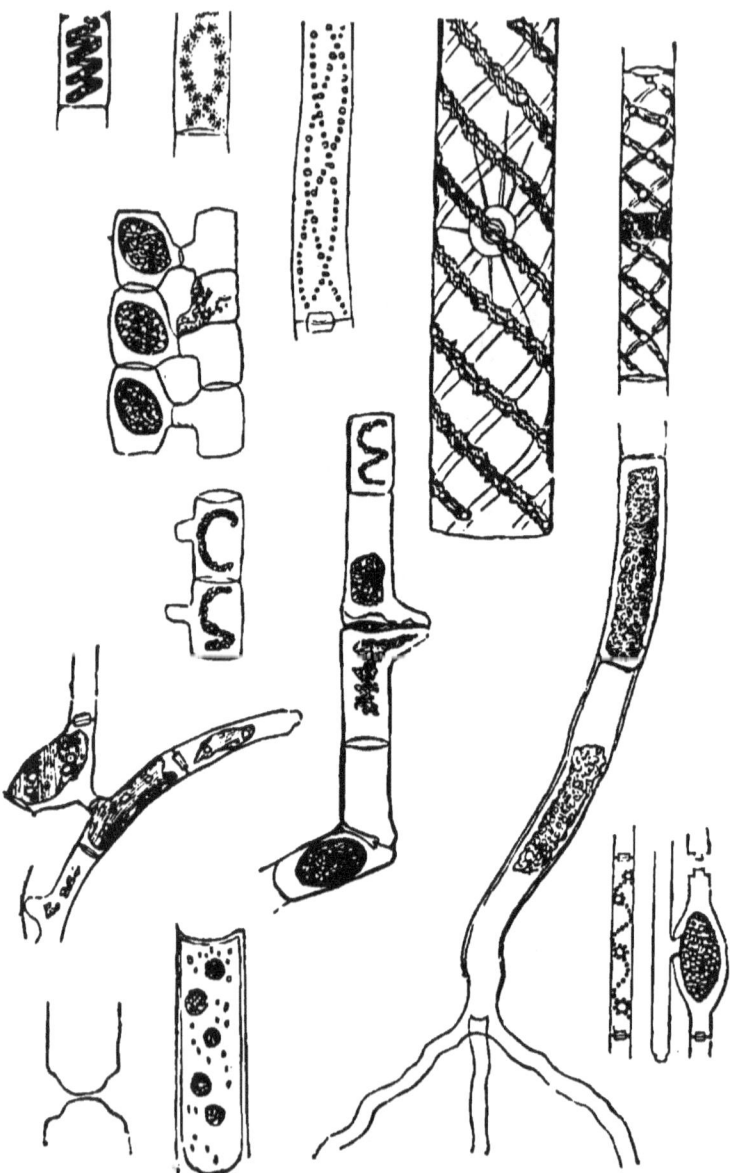

Fig. 351.—Spirogyra in different stages of growth and reproduction (magnified).

MICROSCOPIC PLANTS. 271

will subsequently spring. The transparent wall allows of the movements of certain brown bodies to be visible (probably zoospores). The contents of

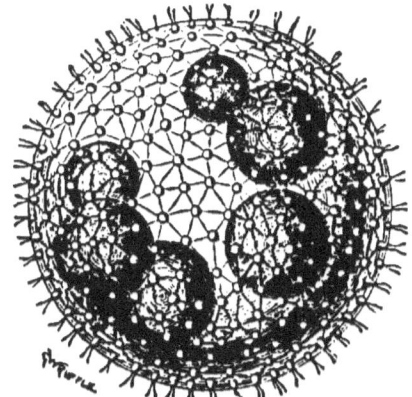

Fig. 352.—*Volvox globator.*

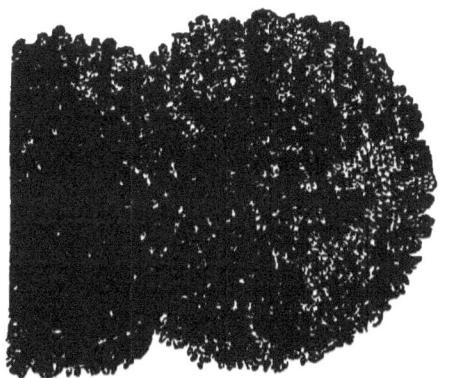

Fig. 353.—*Phyllactidium pulchellum.*

the cells are also seen to change into green zoospores, which escape from the ruptured cell.

A singular and very pretty vegetable form, known

as *Phyllactidium pulchellum*, is often found associated with Volvox in the same pond. In short, it seems to do duty in winter for the absence of Volvox which is then in the resting stage. It is a discoid water-weed, which only requires to be carefully looked for to find it much more abundantly than has been the case. *Volvox globator* is sometimes

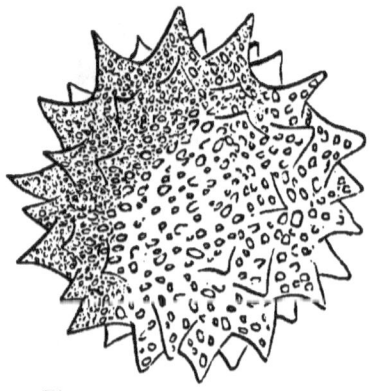

Fig. 354.—*Volvox stellatum*.

found enclosing another species called *Volvox stellatum* (Fig. 354). The latter is believed to be a form of "alternation of generation," not uncommon among the lower groups of life.

The Diatomaceæ have long been special favourites with people who possessed good microscopes. And no wonder. It is a strange sight to see minute canoe-shaped objects like Navicula (real "little ships," as the word means), and Stauroneis moving slowly, sometimes jerkily, amid the microscopic jungle to be seen in any gathering. First they move this way, then that ; and to this very day the *actual* secret of their power of locomotion has not been satisfactorily made out. This ghostly method of locomotion caused the diatoms to be formerly included among Animalculæ. You can hardly go to the wrong place for them, if it is only very damp. Squeeze a handful of moss out of a

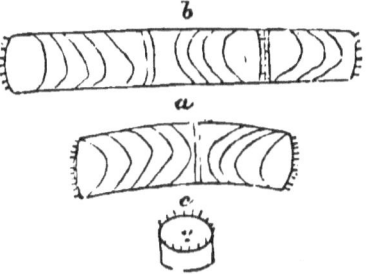

Fig. 355.—*Liparogyra dentreteres.* a, Arcuate frond ; b, straight filament ; c, valve.

hedge-bank, and the drop of water will be almost certain to contain species of diatoms. The same with the Sphagnum, or bog-moss of our mountains, or even the damp walls of caves, etc.

Our ardent band of collectors heard the above remarks (or some of them) as they walked to and from their happy hunting-grounds. It whiled away the time, and made the journeys seem comparatively short, although many miles had been done. After-

T

wards they had found several curious diatoms, such as *Liparogyra dentata, Orthosira Dickii, Nitzschia vivax, Pinnularia borealis,* etc., by merely washing the specimens of bog-moss out.

Not only in their ability to live in salt and brackish water (as well as in fresh) do the diatoms

Fig. 356.—*Orthosira Dressari.*

differ from the desmids, but still more importantly in their structure; nevertheless, the diatoms are only single-celled plants like their *confrères.* But they possess a siliceous frustule—that is, a skin of natural glass, which remains behind long after the

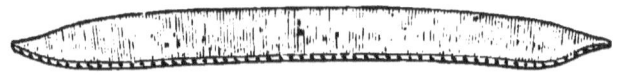

Fig. 357.—*Nitzschia vivax.*

organic matter of the plants is dead. Indeed, it is these accumulated, indestructible frustules or valves which help very largely to form the accumulating black muds of our tidal rivers and estuaries, as well as that along the bottoms of lakes and ponds. The finest "diatomaceous earth" (*kieselguhr*), when saturated with nitroglycerine, becomes the explosive dynamite.

When the desmids die there is an end to them; they leave no trace behind. Not so with the diatoms; their glassy frustules are nearly indestructible. You have only to get a thimbleful of black mud, and place it in a wine-glass; then add a strong solution of sulphuric acid. A very strong smell will be given off, and much effervescence visible. That is a sign the acid is removing the organic matter. Then the mud ought to be washed, and the settlings filtered; then treated with nitric acid, washed again, and so on, until only a little grey powder remains behind. That grey powder will be found to consist chiefly of the flinty shells of diatoms.

Fig. 358.—*Pinnularia borealis.*

Many years ago, M. Deby, the distinguished Belgian microscopist, published a very minute account of how a diatom was constructed, how it managed to secrete its glassy shell, how it split itself so as to form two living individual diatoms where there had previously been only one. "You will find the entire paper in *Science-Gossip* for 1878," said the professor, "translated by Mr. Fred Kitton, who is the best authority on the subject in the world."

How beautifully striated, dotted, ornamented are these microscopical glass cases, intended, somehow, for the use and service of the very lowest orders of plants! For years they have been crucial tests of the best lenses, and many a long and windy article, and many a keen discussion too, has taken

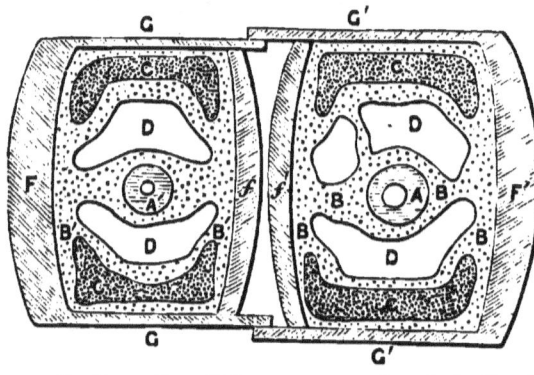

Fig. 359.—Section of a diatom commencing deduplication. A, Nucleus and nucleolus; B, protoplasm; CC, endochrome; FF, valves (highly magnified).

place as to whether the "lines" on certain diatoms were rows of dots or continuous ridges. Even scientific people quarrelled over their differences as political people now do on Home Rule! It will be a long time before the world gets rid of, sloughs off, its inheritance of folly.

Three or four years ago Mr. F. H. Lang drew attention to the exquisitely lovely markings on one diatom, known as *Stictodiscus Californicus* (Fig.

360), found in a deposit in the country to which the specific name alludes. But many of our common and easily found British species are quite as beautifully adorned. Every ornamental pattern is the "Broad-arrow of the Great King, stamped on all the stores of his arsenal."

One can hardly wonder that our distant fore-

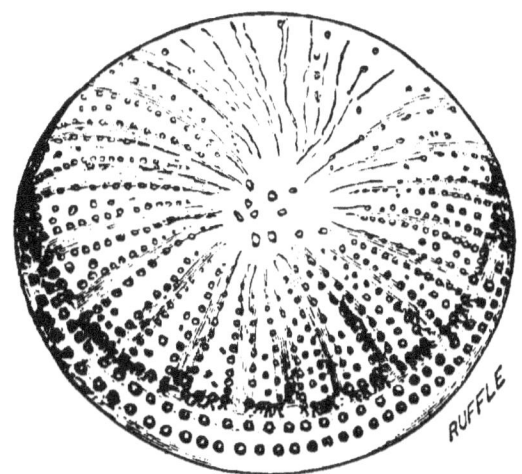

Fig. 360.—*Stictodiscus Californicus.*

fathers associated exquisite loveliness with minute objects. The diminutive fairies were always beautiful, whilst the giants were always ugly—at least, in the story-books. The grace of God ornaments the invisible flinty valve of a diatom or the limey shell of a foraminifer, as it does the possession of a meek and quiet spirit!

Just now there is somewhat of a "rush" against *teleology* — the matter-of-fact, hard, cut-and-dried (and very presumptive) theology of our grandfathers. Let us not condemn them, although the

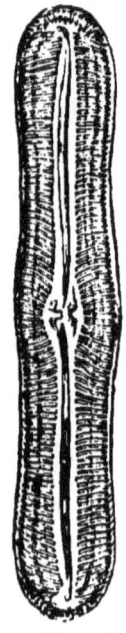 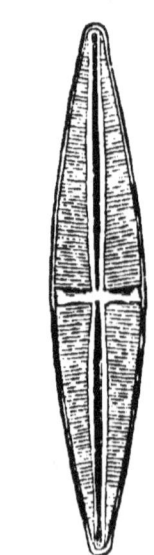

Fig. 361.—*Pinnularia major.*   Fig. 362.—*Stauroneis Phœnicenteron.*

dreadfully conservative spirit of theology has a tendency to glorify that which science condemns. Our forefathers did their best, as honest men, to understand God and His ways. If they did not succeed to our mind, probably they did in spirit. He must be worshipped in spirit and in truth.

MICROSCOPIC PLANTS. 279

They endeavoured to worship Him so, according to their lights. Now that science comes to the help of a reverent man's heart, let him not scorn the

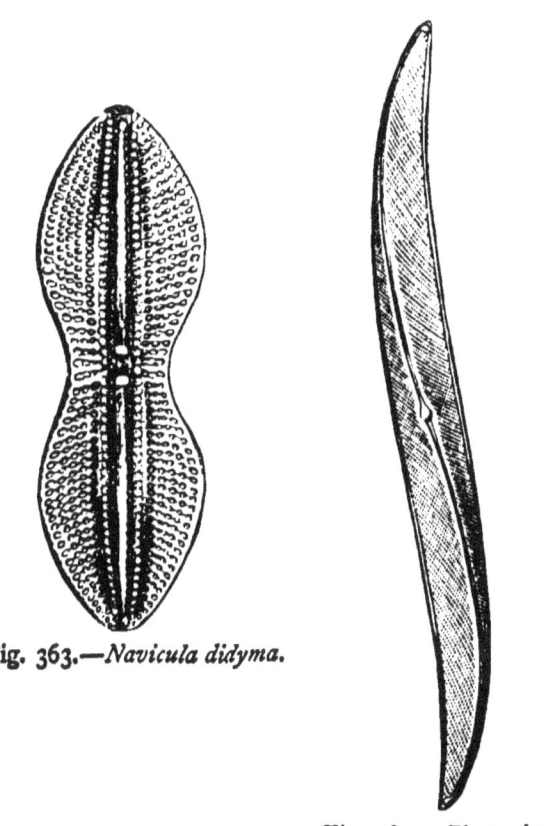

Fig. 363.—*Navicula didyma.*

Fig. 364.—*Pleurasigma formosum.*

day of smaller and feebler and even more bitter things.

The older *teleology* is gone, practically dead, and

almost buried. It was too presumptive. It failed because the child wanted to lay down the rules of government, and dictate to its Wise Father, and teach Him how to rule His own world, although it knew nothing of the infinite battalions of suns and planets. Science doesn't know everything, even now. Indeed, the fear is lest modern academic science should usurp the theological chair, and prove doubly dogmatic. There are men who would dethrone St. Paul, and put up St. Darwin in the vacant seat. Neither St. Darwin nor St. Paul would assent to the change.

The truest teleology is that of trying to "seek out God, if haply we may find Him!"—in His works, His Word, His people. Who dares say where God is not to be found, when we see He does not think it beneath Him to ornament the frustule of a diatom the five-hundredth part of an inch in length so beautifully that, when the modern "children of Israel" behold it through the microscope for the first time, they immediately think how capitally the ornamentation might be applied to a new kind of jewellery!

Of course, all my young readers know now that I have been employing Jack and Willie and the professor, and the rest of the lot, as conscientious candidates do their "friends" at a parliamentary election—for what use they can make of them. I've now done with Jack and Willie and the professor, and even Jack's sister and cousins; they were only my wax-works!

But possibly my young readers may have learned something more than they knew before, because of my batch of hypothetical friends. If so, what do they want more? If they have eyes, pocket-lenses, microscopes, the same stock of common objects are available to them. *Common* only to the multitude who regard them not; *uncommon* to those who see in every living object, animal or vegetable, macroscopic or microscopic, additional evidences of the Fatherhood of a common God! Verily, the life we live and lead here becomes then only the ante-chamber of the life to come. Its lessons and illustrations are so many side-lights of the lessons we shall learn there. If there has been an unbroken continuity on our planet in geological times from animalcule to man, from the Archæan period up to now, where is the bold sceptic who will declare that the stream of that vital tide shall henceforth be arrested? Who will deny the possibility, at least, of its flowing to a higher level yet through the golden gates beyond?—or who will deny that the total sum of the life of the world, terrestrial and celestial, may be but the wonderful development of a Divine idea, as continuous in its unbroken evolution as that of a bird from the egg?

INDEX.

A

Amœba, 242
Antennæ, 52
Aphides, 105
Arrenurus integrator, 177
—— sinuator, 174
—— tricuspidator, 177
—— truncatellus, 177
Asilus crabroniformis, 95
Asterosporium, 253

B

Bee, leaf-cutter, 89
Birds—
    Blackcap, 42
    Blue tit, 28
    Chaffinch, 29
    Cuckoo, 20
    Dipper, 39
    Goat-sucker, 43
    Jay, 22
    Kingfisher, 24
    Lapwing, 30
    Long-tailed tit, 26
    Reed-bunting, 35
    Sand-martin, 32
    Sedge-warbler, 34
    Whinchat, 30
    Yellow-hammer, 29

Birds' eggs, 46
——, arrangement of, 48
——, preparing, for cabinet, 46
Blackcap, 42
Bluebottles' eggs, 56
Blue tit, 28
Bombylius, 93
Bramble, brand of, 253
Brands of meadow-sweet, bramble, etc., 253
Brown lizard, 199
Bucentes geniculatus, 95
Bulimus acutus, 130
—— obscurus, 129
Butterflies, 51, 52
——, eggs of, 56
——, killing, 81
——, preparing, for cabinet, 81
——, scales of, 55
——, white cabbage, 65

C

Caterpillar of small eggar moth, 68
—— puss moth, 70
—— sphinx moth, 69
—— white cabbage butterfly, 65
Caterpillars, 67
——, miner, 62
Chaffinch, 29

Chitine, 12
Chitinous, 136
Chrysalis of white butterfly, 67
—— of house-fly, 99
Cilia, 140
Clear-wings, 53
Closterium, 262
Collecting-bottle, 258
—— for diatoms, 259
Compositæ, 93
Conops, 94
Corethra plumiformis 155
Crown animalcule, 233
Cuckoo, 20
Cyclops, 207

D

Dermestes, 116
Desmids, 261
Diatoms, 272
——, collecting-bottle for, 259
Dipper, 39
Diptera, 90
Dissecting beetles, 146
—— molluscs, 137
Dredging, 141-159
Duck-weed, 210
Dytiscus marginalis, 145
——, parasite of, 147
——, preparing trachea of, for the microscope, 145

E

Eggar moth, small, 68
Eggs, birds', 46-48
—— of bluebottle, 56
—— of butterfly, 57
—— of house-fly, 57, 98

Eggs of moth, 57
—— of Ranatra linearis, 148
Enomostracans, 204
Ephemera, larva of, 150
Epistylis, 246
Euglena viridis, 245

F

Fish moths, 120
—— scales, 11-19
"Fleur-de-luce," 38
Floscule, 233
Flowering rushes, 38
Fresh-water polyzoon, 216
—— sponge, 248
—— ——, ovaries of, 248
Frog-spawn, 185

G

Gall insects, 86
Glow-worms, 105
Gnat, common, 155
——, eggs of, 156
Goat moth, 63
—— sucker, 43
Grasshopper, great green, 117

H

Habitats of insects and plants, 79, 126
Hair-tailed millipede, 115
Hair-worms, 153
Helix caperata, 130
—— nemoralis, 139
Holiday rambles, 83-123

INDEX. 285

Honeydew, 106
House-fly, 197
——, chrysalis of, 99
——, eggs of, 57, 98
——, larva of, 93
Humming-bird hawk moth, 54
Hyalinity, 155
Hyalotheca, 266
Hydrachnidae, 170
Hydra viridis, 212
——, stings of, 215
Hydrophilus piceus, 145, 147
Hygrobatidæ, 170

I

Improvised live-box, 188
—— zoophyte-trough, 202
Infusoria, 235
Insects, habitats of, 79, 126

J

Jay, 22

K

Kingfisher, 24

L

Land-shells, 124, 140
Lapwing, 30
Leaf-cutter bee, 89
Lepidoptera, 50-82
Leptogaster cylindricus, 96
Limnæa, 140
Live box, improvised, 188
—— cell, 202

Lizard, common brown, 199
——, sand, 199
Lizards, 198
Long-tailed tit, 26
——, nest of, 26
Louse-wort, 32

M

Melasmia, 254
Micrasterias rotata, 265
Micro-Lepidoptera, 72
Microscopic plants, 251-282
Millipede, hair-tailed, 115
Mimicry in nature, 52, 72, 93, 95, 117
Miner caterpillars, 62
Mites, among the, 160-178
Moth, clear-wing, 53
——, goat, 63
——, humming-bird hawk, 54
——, puss, caterpillar of, 70
——, sphinx, 69
——, wingless, 72
Moths, killing and preparing, for cabinet, 81
——, scales of, 55
——, sugaring for, 73
Myopa testacea, 95

N

Natterjack toad, 193
Natural History Society, our, 3
Nest of kingfisher, 24
—— long-tailed tit, 26
—— reed-bunting, 35
—— sedge-warbler, 35
—— spider, 111
—— water-ousel, 41

Newt, crested, 190
——, spawn of, 185, 191
——, tadpoles of, 191

O

Odontophores, 135
——, dissecting, 137
Our Natural History Society, 3

P

Paludina, 140
Peewit, 30
Peridinium, 245
Phyllactidium pulchellum, 272
Plant-lice, 105
Plants, habitats of, 79, 126
Plumatella, 216
Podura, 120-122
——, scales of, 121
Polyzoon, freshwater, 216
Puddingpoke, 26

R

Ranatra linearis, 148
——, eggs of, 148
Reed-bunting, 35
Robin-redbreast cushion, 88
Rotifers, 221

S

Sand lizard, 199
—— martin, 32
Scales, butterfly, 55
——, fish, 11-19
——, moth, 55
——, Podura, 121

Sedge-warbler, 34
Shells, land, 124, 140
——, preparing, for cabinet, 132
Sigillaria, 126
Smell, sense of, in insects, 75
Snails, 125
——, jaws of, 135
——, larva of, 140
——, odontophores of, 135
——, preparing, for the cabinet, 133
Spawn, 182
—— of frogs, 185
—— of newts, 185
—— of toads, 185
Spiders, 108
——, nest of, 108
——, spinnerets of, 108
——, water, 111
Spirogyra, 268
Sponge, fresh-water, 248
——, ovaries of, 248
Statoblasts, 218
Stephanoceros eichornii, 226
Stings of hydra, 215
Symbiosis, 167

T

Tadpole of frog, 187
—— of toad, 188
Tetranychus lapidus, 164
—— telarius, 163
—— tiliarius, 164
Toad, natterjack, 193
——, spawn of, 185
——, tadpole of, 188
Trilobites, 152

U

Ulothrix, 268
Umbelliferæ, 93

V

Vaginicola, 245
Viper, 196
Volvox globator, 272
—— stellatum, 272
Vorticella, 245

W

Water-fleas, 204-207
Water mites, 168
—— ousel, 39
Whin, 30
White cabbage butterfly, 65
—— caterpillar of, 65
—— chrysalis of, 67
Whinchat, 30
Wingless moths, 72
"Witches' butter," 254

THE END.

LONDON :
PRINTED BY WILLIAM CLOWES AND SONS, LIMITED,
STAMFORD STREET AND CHARING CROSS.

www.ingramcontent.com/pod-product-compliance
Lightning Source LLC
Chambersburg PA
CBHW032044230426
43672CB00009B/1464